OCCUPATIONAL HEALTH PROBLEMS OF YOUNG WORKERS

INTERNATIONAL LABOUR OFFICE - GENEVA

ISBN 92-2-101051-1

1st impression 1973
2nd impression 1975

CONTENTS

FOREWORD

When questions are asked about the purpose served by the medical supervision of the health of young workers, it is frequently stated in reply that such supervision is valuable for the prevention of tuberculosis and that a medical examination before employment will reveal anomalies or affections such as heart disease or hernia which would be contra-indications for some kinds of employment. Although these aspects of the pre-employment examination and medical supervision of young workers are certainly important, it would be regrettable if the activity of occupational health services were limited to this scope.

Very often, with questions of occupational pathology uppermost in his mind, the industrial physician will see little point in examining and keeping a check on young workers who in principle are in good health. Sometimes the problems that do exist may go unrecognised by him, especially since young persons very often require sympathetic encouragement before they will pluck up the courage to speak of their personal problems and difficulties. He may even begin to doubt the usefulness of the medical supervision of young persons; in that case, he will soon become bored with these examinations and will tend to rush through them as quickly as possible.

It thus appears necessary to bring out more clearly the importance of the medical supervision of young workers and to lay emphasis on the part that the occupational health specialist must play in helping to protect the health of young persons. Attention also requires to be drawn to the multitude of problems which may concern these specialists and become focal points of interest for them, stressing their diversity and complexity. This work, however, is not intended to provide a detailed study of all these problems, but tries to explain how and why the medical supervision of young workers can become one of the most important and interesting tasks of an occupational health service.

The aim of this work, therefore, is to alert employers, workers and the organisations representing them to the health problems posed by young workers and to their significance, while underlining the fact that very frequently a modicum of friendship, consideration and understanding will get better results than purely technical or financial outlays, even quite substantial.

One can hardly over-emphasise the importance for those who are responsible for helping young persons to take their place in our rapidly changing world of a full acquaintance with the nature, characteristics, needs and problems of adolescent human beings.

The doctor can supply information on the significance of adolescence, and the occupational health specialist, in particular, can supply information concerning the implications of the employment of adolescents.

At first glance, the variety of the questions touched on may surprise the reader, who may not immediately grasp their connection with occupational health. But he will soon realise to what an extent the problems of young persons are interrelated, and how essential is a due appreciation of their complexity and fundamental unity for the establishment of the necessary effective and continuing co-operation between all the different people, services, administrations and institutions concerned with young persons either from the medical standpoint or from the standpoint of education and vocational training, either at work or in the community at large.

Although in this work an attempt is made to cover all the health and medical problems of concern to occupational health specialists, it does not deal with diagnosis or treatment, which must be left to individual physicians.

While some readers may regret the absence of any reference to certain noteworthy national laws and regulations or practical achievements, to have attempted any such coverage would have been to run the risk of being incomplete. It has accordingly seemed preferable to refer in particular to the international instruments adopted by the ILO, which have the advantage of reflecting the substance and "common denominators" of the majority of existing laws and regulations. A study of these instruments provides an idea of the general tendencies and policies reflecting the current aspirations of most countries throughout the world.

The ILO wishes to thank Professor Sven Forssman and Dr. Georges-Henri Coppée for the meticulous care and professional competence with which they have prepared this vivid and up-to-date survey of the problems of young workers. Professor Forssman, with the co-operation of Dr. Max Sallstrand and of Mr. Bengt Gustavsson, has dealt in particular with certain aspects of the pathology and psychology of young persons, as well as with trends in national and international rules and regulations relating to the protection of young workers. Dr. Coppée, for his part, has dealt with the problems of young persons with especial reference to the difficulties bound up with their physical and mental growth and development and to the emergence of their personality, as well as to the role of the occupational health specialist in relation to young workers and his activity at the level of the undertaking.

INTRODUCTION

YOUNG PERSONS IN THE WORLD OF TODAY

Increasing attention is being paid in many countries and international organisations to the problems of young persons, owing to the growing importance of the latter in the world's population, and the economic consequences of this trend. It has been estimated that in 1969, 54 per cent of the world's population consisted of persons under 25 years of age. In some developing countries of Asia, Africa and Latin America, youths less than 20 years old account for as much as 50 or even 55 per cent of the population; in the industrialised countries, the figure is approximately 30 per cent. Some 9 to 10 per cent of the world's population consists of youths between 15 and 20 years old.

In some countries undergoing a "population explosion", the happy sight of many children playing and laughing inevitably provokes the thought that all too soon they will have grown into young persons who will be wanting to learn a trade and seeking the jobs which should be theirs, whereas the local state of economic development is such that nothing will come their way unless new solutions are found and new methods applied with all due speed. Already, some developing countries are unable to provide adequate vocational training to many young persons, or useful employment to those who are old enough to work. There are simply not enough productive jobs going for the nation's youth, and this situation is one of the most serious problems confronting these countries and the international community at large.

In the industrialised countries, the problem of youth also exists, but in another form. Here, the increasingly complex organisation of modern society places obstacles in the way of young persons entering adult life, and imposes pre-requirements for the attainment of responsibility. In some cases, it is not easy to divine whether the measures taken to protect young persons - very often without consulting them - are measures taken by society to protect and promote its greatest asset, or to curb a disruptive force which might get out of hand. Yet young persons everywhere have made clear their desire to participate directly in national affairs and have not hesitated to query the established values of society, with its organisations and institutions set up sometimes for them and sometimes against them, but indubitably without their participation. They find that they are treated as children, whereas they regard themselves as young adults.

In both the developing and the industrialised countries, it is necessary first to define and identify the problems of young persons. The solutions to these problems must be determined not in isolation but within the broad context of national economic, social and cultural development. They must be decided upon with the participation of young persons, who must be granted new responsibilities in elaborating the policies and programmes designed to promote the development of their country from the standpoint of economic growth and of the mobilisation and rational utilisation of human resources.

The distribution of the juvenile population between urban and rural areas varies considerably from one country to another. Over-all, more than 75 per cent of the world's youth is living and working in rural surroundings. A large proportion of young persons are engaged in agricultural work in farms or plantations. They tend to be excluded from large modern industrial undertakings, where conditions are generally better, and are often employed in small artisanal, industrial or family concerns, as domestic servants, and in a variety of minor trades in town or in the countryside, that is to say, in sectors where laws and regulations are difficult to enforce and control and inspection is precarious or non-existent. Moreover, existing laws and regulations are most usually aimed at heavy industry and mining, whereas the needs of young persons employed on the land and in the least-privileged economic branches are often neglected.

A comparison between living conditions in towns and in the countryside rarely turns out to the advantage of the latter; earnings are low, and holidays, as well as opportunities for education, promotion or recreation are limited. Consequently, many young persons leave the land to seek their fortunes in the towns, even if the prospects there are minimal. In the towns, they frequently experience the greatest difficulty in finding work, are likely to be exploited, and are constantly threatened by poverty and disease.

In the industrial sectors, the problems are different. Technical progress and changes in working methods and in the organisation of work call for increasing knowledge and specialisation. The time required to acquire this knowledge will delay the access of the young worker to active employment. Young persons who have not undergone any serious vocational training, either for lack of an opportunity or because they were in a hurry to start earning their living, stand little chance of obtaining interesting work or substantial promotion. In addition, a young worker without sufficient formal training will have difficulty in keeping pace with technical progress and will soon find himself under a handicap.

On the other hand, the young worker may find that the fact of obtaining certain kinds of qualifications does not bring with it the social status and prestige for which he had hoped, and he will be disappointed to find that the technician in today's industry is the equivalent of yesterday's skilled worker.

If the international Conventions on the subject were universally ratified and strictly applied, child labour would be abolished. In actual fact, millions of children who should be at school or at play are at work, sometimes before their seventh or eighth birthday. In many countries, young persons start work between the age of 12 and 14, that is to say, before they are physically and mentally ripe to do so. The plague of child labour (in several countries, children account for up to 10 per cent of the working population) is still an extremely serious problem, all the more so since the occupations concerned and the conditions of work involved are often unhealthy and pose a very real threat to the physical and mental development and welfare of the children concerned.

Unfortunately, it is a plague which is very difficult to stamp out, for it has its roots in poverty and a lack of schools. So long as the family income is insufficient and the work of the children helps the family to obtain the bare necessities of life, it will not be possible to abolish child labour. On the other hand, where schooling is not provided at all, or is provided only for a few years, it will be better for these children to make themselves as useful as they can to their families rather than be left to while away their time in idleness. Thus the question of the age at which the young person starts working is directly related to questions of economic and social development.

CHAPTER I

ADOLESCENTS AND THEIR DEVELOPMENT

Physical Development

Puberty

The start of adolescence may be said to coincide with puberty. Etymologically speaking, this word refers to the growth of hair in the pubic and other regions, but it is habitually used to denote the appearance of the procreative faculty; it marks the end of childhood and the beginning of adolescence. Whereas in girls the transition is sharply defined (first menstrual cycle and first menstruation), this is not the case in boys.

The age of puberty may vary, depending on the geographical region, mode of life and, in particular, the individual concerned. It was formerly accepted that puberty occurred earlier in hot countries (average age: 8 to 10 years for girls, 10 to 12 years for boys) and later in temperate countries (average age: 12 to 13 years for girls, 14 to 15 years for boys), but it may be wondered to what extent this idea was founded on precise findings. At present, for example, the age of puberty noted by the school's medical service of a State in the Arabian Gulf (tropical zone) is the same as in Europe, namely, 12-13 years for girls and 14-15 years for boys. It would seem that diet, the way of life, the stimulation of sexual awareness by the environment and social and economic factors play a more active part in determining the age of puberty than do the racial or climatic factors which hitherto were taken mainly into consideration.

It should also be noted that the age of puberty appears to be subject to gradual change, and that in Europe, for instance, girls are now beginning to menstruate earlier than in the past.

The appearance of puberty is a consequence of complex hormonal changes which have been initiated during the pre-puberal period. These changes are also responsible in large measure for the changes noted in height and weight, as well as for the appearance of secondary sexual characteristics, with all the attendant consequences on the emotional plane.

Increase in Height and Weight

A rapid increase in height occurs during the year preceding puberty. Marked growth occurs during the period from 10 or 11 to 14 years of age in girls and from 12 to 16 years in boys. Maximum height is reached at around the age of 18 years in girls and 20 years in boys. Before puberty, the increase in height is due mainly to lengthening of the lower limbs, while after puberty it is accounted for by the lengthening of the trunk. From the start of puberty onwards, differences in height between boys and girls and between individuals become progressively more marked.

A rapid increase in weight takes place at around the time of puberty, the average periods in which a substantial weight increase is noted being about a year later than those just quoted for height increase. Normal adult weight is reached towards the age of 20 years.

Significant statistical variations in height have taken place over the centuries. In Europe, the average height of the population has been steadily increasing over the last century or two. In Japan, a similar tendency can now be noted, but studies based on the examination of ancient garments, coats of arms and weapons carefully preserved until the present day suggest that several centuries ago the average height of the population was greater still. Thereafter it decreased, and is now increasing again. Various theories have been put forward to explain these variations, none of which is wholly satisfactory; several factors must be taken into account, in particular nutrition and social and economic considerations.

In order to assess the physical growth of an adolescent, reference is generally made to graphs and tables which give mean statistical values for bodily height and weight at various ages. Such documents must be relevant for the population studied and sufficiently up to date. It must be borne in mind that the figures given in these tables are average values; therefore, an adolescent may fall substantially short of or exceed them and still be perfectly normal. Account must also be taken of the wide variations between individuals and of the fact that growth tends to take place in successive steps; thus an adolescent who is near the lower limit may well exceed the average value two years later. It should also be remembered that it is more important to assess height and weight with respect to physiological age (development of puberty and of bone structure) than simply with respect to calendar age. Nor should it be forgotten that increases in height and weight do not necessarily run parallel, and that more or less substantial divergencies may be found at certain stages. Thus one must avoid coming to conclusions on the strength of an isolated finding, but should follow a person's development step by step.

Morphological and Anthropometric Changes

A sexual dimorphism appears in girls at around the age of 15 years and in boys at about 17 years, or occasionally earlier; in boys, the widening of the biacromial diameter exceeds that of the bitrochanteric diameter, whereas on the contrary, in girls, the widening of the pelvis predominates. Until the age of 15, the fat distribution is substantially similar in both sexes; in adolescence, however, the tendency for fat to concentrate in the lower portion of the trunk becomes more marked in girls and less marked in boys, in whom the fat tends to be distributed in the upper half of the body.

The biotype changes; modifications occur in a whole series of bodily proportions, such as, for example, the ratio of the length of the lower limbs and of the trunk, as well as that of skull and face size. It is possible to measure and study a large number of biometric variables, to calculate indices, ratios, and so on. Although the value of such studies must not be disregarded, one should also not read into these biometric data more than they actually contain. For example, vital capacity is a biometric measurement like any other, and while it may be of great interest to follow variations in this parameter, it should not be given undue priority in the assessment of aptitude for effort.

Individual variations become more marked, as does the biotype determined by genetic or racial factors. The study of certain biometric data may find interesting practical applications in the ergonomic design of workplaces.

Secondary Sexual Characteristics

The appearance of the secondary sexual characteristics is, of course, by definition a characteristic of puberty. To some extent, their study will provide an idea of the endocrine balance and its evolution in time. It is therefore important that the medical examination of adolescents should follow up their development and that this should be appropriately recorded.

Organic and Functional Development

Needless to say, all the organs (heart, lungs, bones, muscles, etc.) continue to grow, and the different functions (cardiovascular, respiratory, muscular, etc.) continue to develop towards the adult state. However, this evolution does not always take place simultaneously, and dissociations are frequently to be observed, which must be taken into account; these questions will be treated later on.

Affective Consequences of Puberty

In studying problems of growth and development towards adulthood, their repercussions on the person concerned must also be considered. Adolescents are naturally prone to ask themselves many questions about the changes that they are undergoing (increase in size and weight and development of secondary sexual characteristics). In some cases these changes are eagerly awaited, but in others they may be resented. The adolescent wants to be normal, and it is very difficult to persuade him to accept the existence of variations within normality. Thus, for example, one should avoid telling an adolescent that he is fairly small or that his height and weight or his sexual development are slightly below the average, for this will be directly interpreted as meaning that he is abnormally small, that his puberty is abnormal and that he will never become a normal adult.

The most important thing is to get the young person to accept himself as he is. It matters little whether an adolescent whom the doctor has examined and found to be normal is tall or short, fat or thin for his age or whether his sexual development is at an advanced stage or not; what does matter is that he should come to terms with himself, and it is necessary to encourage this self-acceptance, which is far from spontaneous.

It is also necessary to satisfy the thirst for information and enlightenment displayed by adolescents as regards the facts of their sexual development. Most frequently, an adolescent will not consult a doctor merely to find out whether he is "normal", but on the occasion of a preventive medical examination he will be very glad to have an opportunity of allaying any doubts by speaking freely to the doctor; however, he will do this only to the extent that he feels that the doctor is sympathetically inclined towards his problems.

At the pre-employment medical examination, the industrial physician should note any signs of retarded growth or puberty; the periodical examinations will give him an opportunity of uncovering abnormalities of development. The causes of these conditions should be pinpointed and the possible diagnosis established by the regular physician or medical services, to whom the person concerned or his parents should be directed without fail.

Mental Development

General Characteristics

If one adopts Sheldon's definition of personality as the dynamic organisation of the cognitive, affective, volitional, psychological and morphological aspects of the individual, it may be said that adolescence is one of the crucial periods of life for the development of the personality. During this period, a series of changes take place which may be summarised as follows:

- intellectual progress, with specialisation of the intelligence;
- increasing affective matureness, with development of sensitivity and imagination; this is expressed principally in the appearance of differentiated relationships between boys and girls, as well as of idealism and of the need to commune with other persons and for self-fulfilment;
- organisation of the personality, with affirmation of the self, expressed particularly in individualisation, the personality crisis of youth and the desire for independence;
- discovery of the world of values, the world of ideas and of the mind, and of human values.

These changes are integrated in an over-all dynamic process, involving countless interactions. For development to be harmonious, all these aspects of the personality must progress; if one of them is deficient, the others will be retarded. The personality of evolving adolescents is fundamentally volatile and prone to change with time. Today's findings hardly provide a reliable clue to what may happen several months or years hence.

While the hormonal changes in puberty certainly play an important part in stimulating the mental development of adolescents, it should not be concluded that the latter is brought about exclusively by endocrine changes, and that it is entirely dependent on somatic puberty. In fact, independent mental development *per se* does take place; the attendant processes often keep pace with somatic development, but do not depend on it.

It must also be noted that the evolution, rate and conditions of development are prone to wide individual variations, and may also vary depending on the education and training received and on the family, school, occupational and social background, as well as on the standards and values of the society in which the adolescent is living. At present, there is a tendency for young persons to acquire a measure of maturity at an increasingly early age, and it is not unusual to find them intervening actively and effectively in fields which had previously been regarded as strictly adult preserves, such as politics, trade unionism, economic development and social questions.

The occupational health specialist must always remember that although adolescents have their physical needs and problems, their mental needs and problems are much greater. Through his advice and actions, he is in a position to assist in safeguarding and promoting the mental health of young workers.

Development of Intelligence

In the course of adolescence, intelligence undergoes substantial changes, both quantitatively and qualitatively. The intellectual faculties, representing the ability to understand, to reason and to acquire knowledge, increase regularly until 15 years of age on average. They can be assessed to some extent by tests. In the usual psychometric scale, the performances achieved between 14 and 16 years of age are regarded as maximum values which will be maintained in adulthood. As the mental faculties develop, they become diversified and specialised; they do not develop in the same way in all fields, so that a possibility of progress still remains after the age of 15 in particular skills, such as mathematical intelligence or visualisation in space. Knowledge, which is acquired, may well continue to increase throughout life.

Between the ages of 12 and 14 years, the mode of thinking changes and there appears an aptitude for abstract thought, for making deductions and for logical reasoning based not only on concrete data and real experience as hitherto, but on propositions, verbal data and symbols.

If the appearance of this ability to think in the abstract is stimulated by education, and its use is encouraged by his environment, the adolescent will quite naturally be inclined to exercise it extensively. This may explain, at least in part, the fondness of adolescents for abstract speculation. Affective changes, which will be considered later, stimulate interest in general problems and broad concepts. Towards the age of 16, adolescents become involved in mental speculation about theories and systems, as well as about philosophical questions, this is the period of initiation into politics, trade unionism and social problems in general. It takes some time for the young person to adjust his thinking to the facts of life; the confrontation of ideas and experience leads naturally to a properly balanced view of things and to the intellectual maturity of adulthood.

The intelligence is open to development in accordance with the tendencies that exist in adolescents, but these are mere potentialities which will be realised only if action is taken towards that end. Interest in abstract questions and spiritual matters is not usually spontaneously predominant, and will be easily stifled by other more immediate preoccupations, or may be prevented from developing by mental or physical overstress or by poor living conditions, or simply because of a lack of interest stemming from immaturity, from personality deficiencies or from a systematic opposition to the adult world.

Feminine and masculine intelligence tend to differ in certain aspects. Thus, the dispersion of intelligence quotient values seems less marked in girls than in boys, girls seem to mature more quickly, they do better in tests with a strongly verbal bias, etc. But these are only statistical differences. Girls frequently evince a certain docility which contributes to good school results but is not necessarily related to intelligence; characteristics bound up with feminine psychology and with the emotional climate peculiar to young girls are also to be observed.

Irrational Mental Life: Affectivity, Sensitivity, Imagination

Affectivity is an essential aspect of personality; it is the capacity to experience sentiments and emotions. Affectivity makes it possible for a person to become interested in things and in other persons. It plays an indispensable part in the development of the personality of which it constitutes the dynamic aspect. A person who loses the ability to experience emotions rapidly loses interest in his surroundings and in outer events, he opts out of social life, loses all sense of motivation and of an aim in existence, and may become a social misfit.

Affective development is as important for mental life as is the course of puberty for somatic life. During adolescence, the capacity for emotion changes and matures. An infantile, essentially selfish and acquisitive personality undergoes, particularly at around 15 years of age, a reorientation towards the outer world, accompanied by a marked increase in the capacity for sympathy, friendship, generosity, co-operation and self-sacrifice. But it must be admitted that this trend is expressed more frequently in affirmations and declarations of principle than in changes in attitude in everyday life, especially family life. Moreover, underneath this layer of altruism, the adolescent remains fundamentally taken up with himself, his physical and intellectual development, his feelings, ideas and problems. Adolescents tend to see everything in relation to themselves, and their point of view is often highly subjective.

Imagination, the ability to form a mental picture of things and inner life as a whole undergo very substantial development during adolescence. At about 14 years of age, a decided tendency to introspection and to sentimental or melancholy day-dreaming may be observed. The flowering of an inner life is essential for the development of the personality. In order for an adolescent to derive progressive enrichment from external sources, not only must he be receptive and interested in the world around him, but the information received must also have an affective content which will react, as it were, with his personality. By using his imagination, the adolescent must be able to analyse, to classify, to compare, to assemble or to distinguish between the external messages received and to imbue them with a varied emotional content in all kinds of ways.

Traits of character will become more marked, and personal affective tendencies will appear. Affectivity not only develops but becomes diversified. Girls will increasingly display typical feminine characteristics such as emotivity, patience, passivity, compassion, narcissism and attentiveness to the opinions cf others, while boys will evince such masculine characteristics as obstinacy, combativeness and the will to gain one's way and to dominate.

The tendency of the adolescent imagination to develop and of the adolescent affectivity and sensitivity to mature will be fully realised, however, only if circumstances are favourable tc such an inner ripening. These tendencies are not spontaneously overriding; more immediate preoccupations, as well as the search for easy gratification, may well sterilise all inner life, and such cases are not uncommon. The development achieved will depend on individual characteristics, on the social, family and physical surroundings, on the education received and on the extent of previous experience. The capacity to become interested in things and to feel must be inculcated in adolescents, their imagination must be stimulated, they must be given an affective culture. Far too little attention is paid to this problem, either in the family or by educators, whose main objective appears to be the accumulation of knowledge. And yet adolescents have vast requirements in these respects, which must be met.

Individualisation, Affirmation of the Personality, Need for Independence

In early adolescence, and especially at around 14 years of age, there occurs a period of uncertainty, timidity, anxiety and uneasiness, which most usually remain unexpressed. Uneasiness about the physical changes he is undergoing, timidity and anxiety about the new relations in process of establishment with his former female playmates and with girls in general, and uncertainty as a result of changes in his relationship with parents and adults, all contribute to a feeling of indefinable malaise. The adolescent feels isolated and misunderstood, and believes that his difficulties are unique.

After this first period of confusion, the adolescent will progressively take stock of his personality and his individuality by defining and affirming them in the face of others, principally adults. He will affirm his individuality and independence in the most varied fields of daily or social life. Towards this end, he may systematically adopt a different attitude to that of the rest of the community, and may select standards, values and modes of behaviour which are intended to stress his individuality, but also bring him closer to other, slightly older young persons. Eccentric clothes, the acquisition of motor vehicles of various kinds, "wild" parties and opposition to the social order, mostly verbal but sometimes translated into acts, are symptoms of this dual wish - the wish to be "different", but also to belong to the social class of the "under 20s", the prestige of which is constantly enhanced by powerful communication media. Although this behaviour has always existed in young persons, its exploitation for commercial ends is peculiar to present-day society.

Adolescents have an urge to live more intensely, more sincerely, more effectively, more enthusiastically and more freely than their seniors. Some of them become passionately interested in problems of culture and of the human condition, towards which they seek to adopt a totally free approach; they like to accumulate objective experiences, but integrate them with a degree of subjectivity which can be surprising.

They are also much taken up with their impressions, their feelings, and the way in which they see the world and the people in it. This is the period of intimate diaries, endless conversations about their respective personalities, and attempts at largely romantic writing or art.

The process of individualisation can sometimes lead to a real crisis of juvenile originality. The adolescent is not satisfied with affirming his moral and psychological autonomy, he wants these to be reorganised, he wants his independence. In the choice of his pursuits and amusements and in decisions about his future life and occupation, although he may not disregard outside advice altogether, he is anxious at least to have the last word. This situation sometimes leads to conflicts, which will be the more serious, the more the adolescent is materially dependent on his family background.

Adolescents want one to be interested in their work, to appreciate their efforts and, if their work is well done, to say so. They want to be useful and productive workers and to be accepted as such and not regarded as children who are there merely to observe, to receive lessons and to be the servants of their seniors.

Very often, adolescents want to be enthusiastic about their work; they are ready to make all the necessary personal sacrifices and to shoulder responsibilities in order to achieve the aim that they have fixed for themselves. Although this tendency is not restricted to adolescence, it is particularly frequent and intense during that period of life. This enthusiasm deserves to be encouraged and developed. It must not be damped by cumbersome administrative procedures or measures which effectively prevent young persons from translating their will to create and to serve into action. Nor must it be met with the general indifference or even hostility of adults who have already made their way in life. Yet, alas, this is all too frequently the case.

If this natural tendency of adolescents to strive for ambitious targets is not developed, they will only naturally end by choosing the easy way out, and in such circumstances it is not surprising to see increasing numbers of young persons, for lack of any better ideal, becoming old before their time, devoid of any individual drive, concentrating not on personal effort but on demands, evading any kind of responsibility and displaying no personal initiative whatsoever. These young persons will hold the community responsible for their frustrations, and will wait, without doing anything, for the community to bring about an improvement in their situation. Moreover, it is not astonishing either if young persons sometimes react violently, brutally or explosively in systematic opposition to the irresponsible, paralysing and impersonal existence which, too often, is all that is offered them.

Without going to extremes, one must admit that it is not astonishing that young people should query the value of what is proposed to them and feel concerned about what life has in store for them. One might even add that it is fortunate for humanity that they should do so, and express the hope that some new ideas will result.

Yet while on the one hand adolescents have a need for self-expression and independence, on the other hand they stand in equally great need of discipline and of the presence of authority. They must not be given their freedom outright, but must earn it little by little in order to appreciate it at its full value. An adolescent faced with quite unrestricted freedom to choose between various alternatives will experience great difficulty in deciding, and this situation may strengthen any feeling of anxiety about approaching adulthood, or trigger off systematically negative reactions.

Adolescents stand in need of an authority, of a discipline by which they may be guided as long as they lack the capacity to decide for themselves, or which they can contradict, for such contradictions have a formative value and represent part of the apprenticeship of freedom. It should be pointed out that the lack of authority and discipline and the failure of the family and of society to cope in this respect stem more frequently from indifference than from an excess of goodwill.

The Love of Risks in Adolescents

Adolescents are particularly fond of taking risks; this taste is bound up with their need to express and to surpass themselves. This tendency should not be systematically condemned, nor should it be brutally repressed by dint of multiple and repeated prohibitions, for the result would be the opposite of that which was intended, with a good chance of encouraging the systematic search for useless and spectacular risks. Moreoever, an adolescent who is constantly over-protected will remain ignorant, defenceless and without reactions in the dangerous situations that he will not fail to encounter at some stage and which no protection, however perfect, can hope to eliminate entirely.

This liking for risks in adolescents must be guided and directed towards a worth-while objective. They must be taught carefully to evaluate the risks that can or cannot be taken and also to choose the least risky method of achieving the desired aim with maximum efficiency and certainty.

Although young persons may sometimes deliberately take pointless risks, they also know perfectly well how to exercise great common sense and caution, and to subject themselves to very strict discipline - sometimes better than their seniors - in practising certain hazardous sports such as motor racing, SCUBA diving, parachuting, mountain climbing or potholing.

In the world of work, many cases can also be observed in which prohibitions which are imposed too systematically or without giving reasons for them, or over-protection, produce results contrary to those expected, and may lead to the taking of unnecessary risks. On the other hand, young workers can often be led more easily than their elders to establish and keep up a high level of safe and healthy behaviour in an undertaking, at least when one has taken the trouble to give them the necessary information, explanations and training.

The Adolescent and the World of Values

The values that a child knows and respects are those in which he is drilled; they have a concrete and compulsory character and are subject to rewards or sanctions. In adolescence, the field of values broadens considerably. They assume an abstract nature and are no longer proposed or imposed by persons close to him, but are derived from a much wider range of sources (and personal experiences). They are no longer compulsory, and the adolescent awakens to the concept of the relative nature of values, which lose their absolute character; consequently, they are called into question again and again, and the possibility arises of conflicts at the level of the family or of society. The adolescent wants to establish his own scale of values, rejects those which are thought to be imposed from outside, and searches for an internal discipline.

To establish a scale of values presupposes on the one hand that these values are known, and on the hand that they are judged and evaluated according to certain criteria. The normal tendency of young persons is to look beyond themselves and strive towards lofty aims, and they should not be fobbed off with the goal of easy living in a consumer society, as is too often done for reasons which are mainly commercial and economic. They must be helped to choose objectives which are worthy of them. They must not be spared all effort, but must be helped to mould their personality and character.

When the time comes to choose a profession or a trade, it is important that the adolescent should have been prepared to make his choice as a function of his personal aspirations, rather than of social and cultural conditions. He will require guidance in order to ensure that his wishes can be reconciled with the realities of life, and that his choice is a realistic one, taking into account his aptitudes and skills and economic or other local circumstances.

Within the undertaking, care should be taken to encourage the adoption by young workers of an appropriate scale of values, in which the concept of work well done, and done in adequate conditions of safety and health, will figure prominently.

Evolution of Emotional Balance in Adolescents

In childhood, even in early childhood, behaviour can be noted which indicates that the little boy or girl is attracted to his playmates or to adults of the other sex. After the indifference or even the hostility of the period preceding puberty, a substantial increase in the interest shown by young adolescents in the other sex can be noted. This interest coincides with the sexual differentiation of tastes and interests in fields which go well beyond the purely sexual domain; boys take an increasing interest in sports and public life, they imitate their adult male acquaintances and model themselves on them, whereas girls seek female models and take on feminine characteristics. The awakening of sexuality proceeds by stages, with transitional phases which are necessary for the elaboration and acceptance of sexual differentiation. Whereas sexual maturity is experienced by a girl as a definite state, owing to the appearance of menstruation, in boys it does not manifest itself so specifically, and they experience a need to assert it in order to convince themselves of its attainment.

Adolescents are subject in turn to the very specific urges of instinct and to the vaguer influence of a sentimentality which rests content with imagined love or platonic idylls. The attainment of adult sensitivity and matureness is a lengthy process. In adolescent girls, emotivity and sensitivity are highly developed, while sensuality is experienced more vaguely; tenderness is experienced well before sexual desire, whereas in boys the reverse situation holds good.

The appearance of instinctive urges and emotions of sexual origin undermines the equilibrium generally existing at the end of childhood. The adolescent must assimiliate these new facts, and his personality must evolve from that of childhood towards the adult condition. This indispensable evolution is often far from smooth, and is in many respects a source of concern for adolescents.

At times the adolescent will seek to release himself from the state of tension which he feels deeply and which is sometimes manifested by various nervous tics or psychosomatic disorders (nail-biting, stammering, headache, stomache-ache, precordialgia, etc.).

Certain reactive changes are well-nigh specific of adolescence: these are asceticism and intellectualisation. Asceticism places a ban on the instinct of self-expression; it can become pervasive, and is often associated with masochistic tendencies. Intellectualisation represents an attempt to sublimate the instinctive appetites; it is a favourable factor to the extent that it stimulates the faculty of abstract thought, but its exaggeration may sterilise the development of affective values.

Other reactive changes may occur. One may observe in adolescents a tendency to retreat within themselves, to excessive timidity, to frustration, to negative thinking and to evade reality, or, on the contrary, the development of outright opposition to the environment, possibly accompanied by aggressive or impulsive behaviour accounting for the brutality or destructiveness of certain adolescents.

Information, guidance and education are required. They may be given with advantage during the period of latency at the end of childhood, when the sexual instincts are still weak, but the mind is mature enough to benefit from objective sexual education, which should go beyond the physiological plane and deal also with human values. At the same time every adolescent should be encouraged to develop a personality in keeping with his or her nature and future role in society, which are not the same for men as for women.

Variations of Behaviour in Adolescents

Behaviour must not be regarded as a mode of conduct; on the contrary, it corresponds to a variety of modes of conduct, which may sometimes be contradictory. Variations with respect to a basic line of conduct do not represent departures from "normal behaviour", but in fact constitute such behaviour. In adolescence, these variations are particularly frequent, intense and marked; this provides evidence of the extensiveness of the psychological changes under way and the instability noted is a symptom of the intensity of the processes unfolding in the adolescent's inner self. On the contrary, a rigid, unvarying and stereotyped mode of conduct will suggest that the normal psychological development is not taking place. Passivity and mental inertia, more than wide variations in conduct, may point to a pathological organisaticn of the personality. The tendencies to create and to destroy, as well as to create and demolish oneself, which are present in everyone, will occur simultaneously or in succession in a more marked fashion in young persons.

While it is possible to pinpoint certain psychological traits which are common to all adolescents, their importance must not be overestimated. Sight must not be lost of the fact that in every adolescent, behaviour is largely a function of individual character, experiences, childhood, education and environment. Extroverts will tend to be exuberant while introverts will prefer to isolate themselves in silence and dreams.

Although one must not be dismayed by wide variations in conduct, and should refrain from hasty reactions, it sometimes happens that all reasonable limits are greatly exceeded; such cases are instances of pathological conduct which may be symptomatic of disordered psychological development. Such conduct, as a rule, is no more than an extreme form of normal conduct. It is pathological on account of its excessive nature, not of its structure; it tends to become attenuated and should disappear spontaneously with the ending of adolescence.

It must be pointed out that the stricter the family and community standards, the sooner one will start to speak of the "crisis" of adolescence, although adolescents do not change much from one generation to another. The stormiest stages and aspects of adolescent development are highlighted by today's powerful mass media, a fact which tends to encourage extremes of behaviour by youth as a whole.

Development in Relation to Social Life

Adolescents and the Social Environment

The physical and mental changes which occur in adolescence give rise to changes in the relationship between the subject and the environment; his behaviour towards the latter and towards society alters. The adolescent becomes increasingly aware of social realities and attempts to define his place in society and to adapt himself to society.

The affirmation of the personality and the wish for independence lead to the progressive detachment of the adolescent from the family circle, sometimes accompanied by opposition or conflict. The search for self-assertion and the desire for a measure of social esteem may have happy social consequences, by encouraging success in studies, work or sport, or in artistic or cultural activities, etc., but this tendency may also take an unfavourable turn, and be expressed in systematic and unconstructive opposition, the taking of unnecessary risks, delinquency, etc.

The adolescent will seek to identify himself with new models, which will no longer be the father or mother, as in childhood. This search may have a favourable effect or otherwise, depending on the model selected.

The adolescent will seek to invest his affective capacity elsewhere than in the family circle. At around 13 or 14 years of age, typically, adolescents indulge in comradeship and close friendships and band together in groups. Two conflicting tendencies are noticeable, one being the need for social life, for self-expression and for relationships with other young persons of the same age, and the other the need to isolate oneself in meditation, the observation of the outside world and introspection. Most frequently, the adolescent will display both tendencies in turn, depending on his character, sometimes going from one to the other in a single day, thus giving an impression of instability and inconsistent humour. While the fact of being successively and simultaneously attracted and repelled by these two contrasting kinds of relations with the environment may give rise to an impression of uneasiness and even of inner conflict, these successive episodes in which experience is gleaned from the outside world and incorporated in the personality make an important contribution to the adolescent's development.

The maturing of the social personality of an adolescent is accompanied by complex transformations by which he finally becomes adapted to adult life; these will be the longer and more complex, the more advanced is the environment into which the young person must become integrated. During this evolution, the adolescent will adapt his concepts of social life to make them correspond to reality. He will tend to modify his scale of values in order to be able to fit into the environment in which he lives.

It is understandable that the transition from childhood to the adult state was able to take place naturally, without any long preparations or major conflicts, when it was merely a question of evolving from life in the family to life in the village, but nowadays, with our enormous urban centres and the migration from the land to the cities, the problem is quite different; individual persons do not feel very much at ease in these large communities, and a more elaborate social adaptation becomes necessary. In addition, the collapse or disruption of the traditional modes of life, scales of values and social attitudes, as well as the scepticism and confusion which have taken their place, make this adaptation much more difficult. This situation, indeed, may be at the root of the present widespread social instability in which tensions and conflicts are accented.

The prolongation of the period of adolescence which is to be observed reflects the need for longer preparation for adult life, and in some respects it appears to be a consequence of our civilisation. It may have favourable effects inasmuch as it implies a longer period of mental plasticity, leading, if proper advantage is taken of this situation, to the moulding of a more refined personality. On the other hand, there is an imbalance between physical development which makes the adolescent the equivalent of an adult equipped with an adult's instinctive drives, and the prolongation of the condition of social dependence and irresponsibility. It is not surprising that such a situation may lead to conflicts.

Too mature for childhood, not mature enough for adulthood, the adolescent, nevertheless, has the resource of turning tc groups of young persons of his own age who represent communities fitted to his needs, in which he will feel not estranged but secure, safe to go on quietly building up his personality and discovering his individuality. This may represent a necessary transitional stage in his adaptation to collective life.

Adolescents may join groups organised by adults (youth clubs, sports clubs, choirs, folk groups, etc.) or one of the other more or less coherent, organised and durable groups which spring up spontaneously from time to time. He may also choose to adopt the more or less well-defined standards of conduct adopted by the majority of young persons in his country or, for that matter, throughout the world, of which he has learned through various information media. This trend towards uniformity is becoming more and more marked.

Attitude of Adults towards Adolescents

While it is important to consider the behaviour of adolescents with respect to their environment, it is no less important - but is only too rarely done - to take into account the ways in which adults consider adolescents and adolescence. Adolescence is also, and perhaps particularly, a problem for adults. It is difficult for the latter not to react strongly in the face of adolescence, either because they look back on this period of life with nostalgia, or on the contrary because they resent the pretentious claims and lack of realism of young persons, or perhaps, again, because contact with adolescents brings to the surface some old, deep-lying prcblem which they have never overcome.

The attitude of an adult towards adolescents will vary, depending on whether he regards young persons as destructive, revolutionary, dirty, quarrelsome, cynical, lazy and licentious morons or as defenceless, innocent victims of a corrupt society, whose efforts and work are relentlessly exploited, mercilessly exposed to the stress and fatigue of hard and unhealthy work and to the risk of occupational accidents and diseases. Very frequently, the adult will vacillate between these two extremes or make simultaneous use of these two stereotypes, regarding his neighbour's son as a good-for-nothing layabout, while setting his own son up as a victim of unfair examining boards, and extolling the importance of the protection of youth in public. The adolescent is both the "extra mouth to feed" who must be sent out to work as soon as possible, and the son that one is losing.

The attitudes of the social environment are also very frequently determined by similar stereotypes of opposite natures which are evoked successively or even simultaneously, giving rise to a whole series of rather incoherent emotional reactions.

Indeed, one may well speculate about the precise goal of the hotch-potch of measures aiming sometimes at the protection of young persons and sometimes at the protection of adults (against young persons?) which, even in highly-organised and regulated societies, appear to have been decided under the influence of emotional factors rather than as part of a logical whole. There are many administrative and legal provisions which young persons might interpret as being decreed for the protection of adults; they relate, for example, to acquired rights, seniority, career prospects, titularisation in administrations or universities, age-related, not work-related, increases in annual holidays and in remuneration, or insistence on a level of qualification which is not only too high but unrelated to the job. In addition, many youth protection measures are inspired much more by an over-protective sentimentality which regards adolescents as "poor little victims" than by the realities and necessities of life. Young persons do not like to be treated as "hapless children" who must be protected at all costs, any more than they like the numerous prohibitions which are imposed on them "for their own good" but without their advice or consent, and which, in their eyes, represent attempts by adults to thwart their normal and necessary impulses and to keep them prisoners until they become perfect average citizens.

Adolescence, a Dynamic Social Phenomenon

Adolescence, thus, is the period of transition from childhood to adulthood. It begins when the individual ceases to be a child, and it is agreed that this occurs with the onset of puberty. It ends when the individual has reached manhood or womanhood, with the physical and mental development and maturity that this implies, and has acquired the social status of an adult.

The duration of adolescence varies considerably, depending on geographical and environmental factors and also on the varying rates at which the individuals concerned acquire the necessary qualities and adapt themselves to the greater or lesser complexity of the social and cultural environment into which they must perforce fit.

Adolescence must be viewed not as a condition but as a state of balance subject to ceaseless change. It is characterised essentially by processes of growth, development and maturation which proceed stepwise and not always harmoniously.

The expressions "adolescent", "young person", "minor" or "young worker" refer to youths who are no longer "children" but who are not yet regarded as "adults" or even "young adults", who neither enjoy all the rights nor undergo all the obligations of adulthood, and who often form the subject of special protection.

In practice, the term "adolescent" usually refers to girls aged, on average, between 12 and 18 years old and to boys aged, on average, between 14 and 20 years old.

The term "young persons" is vaguer and embraces both adolescents and somewhat older persons, when it is not wished to classify the latter as young adults. One may say that it is habitually applied to girls under 21 and boys under 24 or 25 years of age. It has been noted that the extreme limit of growth phenomena is situated at around these ages, at which it may also be said that the social status of adulthood has been attained.

The term "minor" describes young persons who have not yet acquired all the rights of citizenship and who are subject to special protective measures. Legal majority is frequently set at 21 years, but some emancipation may take place earlier.

The term "young worker" refers to youths of both sexes who are admitted to employment but are covered by special provisions of the labour legislation. Depending on the laws and regulations in force, the age groups coming under this head may run as follows:

- 14 to 18 years, in many countries, and in the first international Conventions covering young workers;

- 14 to 21 years, in many other countries;

- 15 to 19 years in still other countries and in the second (revised) series of international Conventions covering young workers;

- 15 to 21 years, in certain recent international regulations, and in the ILO Model Code of Safety Regulations for Industrial Establishments.

Although chronological age is an important factor, it must not be overlooked that in a group of young persons of the same age, the "physiological age" of individuals may vary substantially. Some adolescents of 14 or 15 years of age, for example, will already have reached an advanced stage of puberty, while in others it may be in full swing and in others still it may hardly have begun. Obviously the characteristics (strength and capacity for effort) and needs (diet, sleep, psychological, etc.) of these persons will differ fundamentally according to the case. In concerning oneself with adolescents or with young workers, one should always remember that one is dealing with a heterogeneous population to which ready-made formulae cannot be applied without making very great allowances for the wide variations and time lags which may exist between calendar age and physiological age (i.e. age as determined by development of puberty, height and weight and bone structure).

Factors Required for Development

Nutrition of Adolescents

Nutrition and dietary habits. In order to grow and gain weight, a supplementary food intake is required. The health and physical resistance of a young person depends largely on the quality of his diet.

Adolescents need more food than adults, not only to maintain growth, but also because their greater spontaneous activity and exuberance calls for an additional intake of calories. For a given amount of work, their energy requirements may be greater than those of adults, because they do not spare themselves and have not acquired the knack of working efficiently with a minimum expenditure of effort.

Thus the importance of ensuring that young persons generally are well fed is obvious. In practice, however, it is frequently found that adolescents merely seek to satisfy their hunger, and are prone to display stereotyped dietary habits, certain foods being consumed systematically over long periods. Moreover, young workers who are starting to earn and able to dispose freely of at least a part of their earnings are bound up with many other matters and will spend their money on many other things than on rich and healthy food.

The diet should be spread out over the day. It seems that the importance of breakfast requires to be stressed; all too often, this is either skipped or reduced to the minimum, in the shape of a cup of coffee or tea (sometimes with the day's first cigarette!) and yet by spending a quarter of an hour over a more adequate meal, the young person would be able to start the day in much better conditions and with a better store of physical and mental energy. Mealtimes should be sufficiently long and the atmosphere should be calm and relaxing. It must be pointed out that meals taken at home, although most desirable in themselves, are not always of a high standard. This may happen because the family income is too low, or, as is often the case, because much of it is spent on material comforts (e.g. television sets, motor vehicles, outings and holidays, etc.); allowance must also be made for the ignorance of parents, and sometimes for the existence of traditions according to which the food requirements of adults take precedence over those of children. In addition, if the mother is working, she will hardly have the time to prepare really nourishing and varied meals.

The teaching of healthy dietary habits should begin in childhood; after the age of 18 or so, bad habits become virtually ingrained. This education is best given in the family circle; young persons who have been taught to eat properly will generally be conscious of the need for a healthy diet.

Calorie intake and metabolism. The daily calorie requirements of adolescents are greater, in relation to body weight, than those of adults. It is also important to preserve a certain balance between proteins, carbohydrates and fats.

In boys, the necessary calorie intake increases with age and becomes stabilised after the age of 18. In figures, taking the standards recommended in France or the United Kingdom, the average requirements for a normally developing adolescent who will reach the weight of 60 kg at about 18 years of age, displays normal physical activity and lives in a temperate climate, is as follows: age 12-13 years, 2,500 kcal; age 14-15 years, 2,800 kcal; age 16 years and over, 3,000 kcal. In 1957, the United Nations Food and Agriculture Organisation (FAO) recommended higher levels, namely, 3,100 kcal between 13 and 15 years and 3,600 kcal between 16 and 19 years; on the other hand, lower levels were recommended in 1968 in the United States. The desirable calorie intake for a young adult man displaying moderate physical activity is set at 3,000 kcal (UK) or 3,200 kcal (FAO).

For normally developing girls (55 kg at 18 years of age) displaying a normal physical activity for their age and living in a temperate climate, the average requirement according to the standards recommended in France or in the United Kingdom is: age 12-13 years, 2,300 kcal; age 14-15 years, 2,300-2,500 kcal; this represents a ceiling after which the values tend to decrease slightly (2,200-2,300 kcal). The 1957 FAO recommendations were higher, namely 2,600 kcal at age 13-15 years and 2,400 kcal at 16-19 years. The desirable calorie intake for a young adult woman displaying normal activity amounts to 2,200 kcal (in the UK) or 2,300 kcal (FAO).

These figures should not be interpreted too restrictively, but regarded above all as useful yardsticks which illustrate the evolution of the calorie requirements during adolescence. Needless to say, individual variations are especially significant; the needs of adolescents of the same age will differ, depending on their relative growth in height and weight and physiological age. While calorie requirements vary with age, they are very largely dependent on physical activity and work performed. This factor is particularly significant in adolescents, whose levels of activity may vary enormously; thus in this group, normal daily calorie intakes can be noted which exceed or fall short of the average intake by as much as 30 per cent.

The influence of climatic factors upon the recommended calorie intake is taken to be the same in adolescents as in adults. If one attributes the value of 100 to the calorie intake required by a young person of a given age and height and weight, living in a climate where the mean annual temperature is 10°C, the calorie requirements will be around 101.5 or 103 for mean temperatures of 5°C or 0°C respectively, and 97.5, 95 or 92.5 if the mean temperatures are 15, 20 or 25°C respectively (standards indicated by FAO).

It should be noted that, speaking very broadly, if adolescents engage in sufficient and normal physical activity for their age, the sensation of hunger and the appetite constitute a faithful guide, leading automatically to the establishment of the necessary balance between energy expenditure and food intake.

The basic metabolism, which is high in childhood, decreases progressively during growth and adolescence, except for the two years before puberty, during which a relative increase is observed. The basic metabolism of adolescents as a function of weight or body area is higher than in adults, the difference being of the order of 12 per cent. A similar evolution is found if one considers the ratio between calorie consumption and body area; whereas the former increases, the latter also does so, but somewhat more, with the result that the ratio between these two variables decreases progressively during adolescence, the decrease being parallel in boys and girls; as a yardstick, it may be said that if the mean calorie requirements per m² of body surface are equal to 2,000 at the start of adolescence, they will fall to 1,500 at the end.

Protein intake. Retention of nitrogenous products occurs during the period before puberty, when growth is rapid and the nitrogen balance is clearly positive. It follows that in adolescents, the optimum daily protein intake in relation to body weight requires to be substantially higher than in adults.

While there is general agreement on the importance of adequate protein intake, evaluations vary substantially. Some authors estimate the protein requirements of adolescents at 2 g/kg/24h, which is equivalent to some 90-120 g of protein per day. The standards published by a joint FAO/WHO Committee in 1965 set the following figures: for adolescents (boys and girls): age 13-15 years, 0.7 (0.56-0.84) g/kg/day; age 16-19 years, 0.64 (0.47-0.71) g/kg/day; for adults, 0.59 (0.47-0.71) g/kg/day.

The protein intake per gramme and per day recommended in the United Kingdom (1969) is as follows: for boys, age 9-12 years (average weight 32 kg), 63 g, age 12-15 years (average weight 45.5 kg), 70 g, age 15-18 years (average weight 61 kg), 75 g, moderately active young adults (average weight 65 kg), 75 g; for girls, age 9-12 years (average weight 48.6 kg), 58 g, age 15-18 years (average rate 56 kg) 58 g, moderately active young adults (average rate 55 kg), 55 g.

Since during growth a substantial intake of essential amino acids (important for synthetic processes) is required, and since these are found chiefly in animal proteins, it follows that the diet must be particularly rich in the latter.

In order to achieve a proper nutritional balance, the protein intake must cover a certain proportion of the energy needs, commonly evaluated at 10 per cent, and occasionally at 14 or 15 per cent. It is also recommended that the protein intake should amount to 25 g per 1,000 kcal consumed per day.

Although these dietary needs may have financial implications, it is particularly important that they should be met, not only within the family, but also by the meals provided in schools, training colleges or factory canteens catering for large numbers of young workers or apprentices.

Carbohydrate and fat intake. Most of the calorie intake will consist of glucides and lipides, but a certain balance must be struck; the calorie intake obtained from the former should be 2 to 3 times that obtained from the latter.

Animal fats (butter, cheese) and vegetable fats (oil, etc.) should preferably be associated in the diet in the proportion of about two-thirds of the former to one-third of the latter. Also, mainly for digestive reasons, a certain balance should be struck between the easily-assimilated sugars and the starches.

In boys, the average daily glucide needs are approximately 325 g at age 12-13 years, 390 g at age 14-15 years and 410 g at age

16 and beyond; the average lipide needs are approximately 90 g at age 12-13 years, 95 g at age 14-15 years and 105 g at age 16 years and beyond. In girls, the average glucide needs are approximately 310 g at age 12-13 years, 320 g at age 14-15 years and 300 g at age 16 and beyond; the average lipide needs are approximately 90 g at age 12-13 years, 95 g at age 14-15 years and 97-98 g at age 16 and beyond.

Mineral and vitamin intake. These must be adequate. Calcium in sufficient quantities is especially necessary during rapid growth, when the calcium balance is markedly positive; during adolescence, calcium retention continues, but at a reduced level. The daily calcium requirement reaches an absolute maximum at this period (700-1,000 mg compared to 500 mg for an adult), although the proportion, considered with respect to body weight, decreases. The relationship between calcium and phosphorus requirements is 1 in adolescents (1.5 in newborn infants and 0.7 in adults).

An adequate iron intake is also very important, particularly in adolescent girls, and should amount to some 10-30 mg daily; the calcium/iron ratio in the diet should be between 50 and 60 to one.

Vitamin intake is at least as important as in adults; most usually, it will be adequately ensured by a healthy, balanced natural diet, without excessive preparation. Special attention should be paid to vitamins in the B group, particularly if the diet is rich in glucides, as well as to vitamins D and C. The recommended daily intake of thiamine (aneurine, vitamin B1) is 0.4 mg per 1,000 kcal consumed. The daily requirements of vitamin D are of the order of 2.5 microgrammes or 100 international units (the requirements may be met by the action of sunlight on the skin; hence the importance of adequate exposure to the sun). Requirements of vitamin C amounts to some 30-100 mg daily for both adolescents and adults (the diet should include fresh fruit).

Meals at home and outside. For various reasons, many young persons take their midday meals or all their meals away from home. Some are prevented from returning home at midday either by distance or because their parents are both working, whilst others are living in boarding schools or other communities.

Refectories in schools, training colleges, universities and, to a lesser extent and more rarely, industrial and commercial undertakings, play an important part in feeding a large section of the youthful population, and accordingly merit special attention.

The question of hygiene should be considered from many aspects: cleanliness of premises, provision of washbasins for customers in the entrance to refectories, appropriate and easily cleaned kitchen equipment, careful dishwashing and rinsing of detergents, a fastidious staff wearing frequently-cleaned working clothes, having their own cloakrooms (with facilities for frequent hand washing and impeccable sanitary conveniences) and subjected to systematic and continuous medical supervision to prevent any possibility of spreading communicable diseases (tuberculosis, staphylococcal, typhoid fever, etc.).

It must be admitted that hardly anything is done to ensure that young workers receive a proper diet at their place of work. Where refectories exist, they are not always used consistently, and the time devoted to eating may be very short, or hot meals may be provided only exceptionally. Nevertheless, works cafeterias can play a big part by supplying a nourishing and well-balanced midday meal to young workers at moderate or low prices. In countries where nutritional standards are low and the home diet of young workers is deficient and ill-balanced, works canteens or cafeterias can help to make up for such deficiencies and to popularise nourishing foodstuffs.

Industrial physicians should be at pains to concern themselves with the nutrition of young workers. They can encourage the establishment of a refectory in the undertaking, explaining that it will pay dividends by improving the health and output of the workers. Where a refectory exists, there should be close co-operation between its management and the medical service, as regards the preparation of meals and the choice of a rational and well-adapted diet. While the managers of such refectories are specialists in cooking or domestic science, they may be lacking in the necessary breadth of training or experience, and the works doctor can fill this gap by stressing in particular the needs of adolescents and of other categories: miners or farm workers should not receive the same diet as clerks or salesgirls, and vice versa. The medical service should provide the necessary guidance and supervision as regards both food preparation and the planning of menus, and health and hygiene conditions as they affect the premises, the equipment and the staff of refectories.

Such action not only has beneficial effects at places of work, but helps to raise the nutritional standards of the general population. There is evidence that its influence can extend to persons other than the workers concerned, for good dietary habits learned at school or at the workplace also tend to spread to the home.

Physical and Sports Activities

Importance for the harmonious development of adolescents. Reference may be made in this respect to the role of mechanical factors which promote the bone-forming processes and the development of strong ligaments; physical activities also play an important part in the development of an efficient and sturdy muscular system, as well as of a respiratory and cardiovascular system capable of easily sustaining effort. It is during adolescence that these capacities are acquired most easily, and after the age of 20 it is very difficult to make good any deficiencies in this respect.

Although the part played by physical and sporting activities in promoting bodily development is obvious, it should not be forgotten that they also have an influence on mental development. The existence of such a relationship is no cause for surprise if one recalls the extent to which the development of the motor and of the mental faculties runs parallel in small children; in adolescents, this simultaneity is no longer so clearly apparent, but it is still present.

The practice of physical education and sports can play an important part in developing the personality (taste for effort, self-control, fair play, etc.). It can also play a part in building up the social personality, by developing a series of qualities which are of assistance in establishing good human relations (sense of collective effort, team work, solidarity, willingness to assume responsibilities). It permits one to acquire group techniques and a capacity for giving and receiving orders and for disciplined action which will frequently stand young workers in good stead in the exercise of their occupation.

The practice of physical education and sports also tends to promote health in general and healthy habits (physical fitness, cleanliness, avoidance or more limited use of tobacco and alcohol).

Physical activity is indispensable for develcpment in adolescence, yet young persons have less and less opportunity for it. Owing to the proliferation of public transport and private vehicles, adolescents no longer walk to school or to work as much as formerly. At home and in society, there is a general tendency to avoid physical work. In the towns, unbuilt land usable as playgrounds is disappearing, and the streets are reserved for motor vehicles. One sees more and more "young sportsmen" whose activity merely consists in going to football matches or watching them on television. This lack of physical activity among young persons, girls in particular, is becoming an increasingly serious problem because of its negative implications for health; in some countries, it is directly responsible for certain pathological conditions, such as obesity in children who spend their days in front of the television screen eating candy, muscular weakness and orthopaedic anomalies, early appearance of degenerative diseases, etc.

The mere encouragement of sports, however, will very often benefit only those who are already spontaneously inclined towards such activities. The important thing is to reach out to all young persons and inculcate a taste for physical effort in them. Physical education has a formative role which must not be underestimated, and should be integrated in general education. Particular and increasing attention should be paid to this problem.

Physical and sporting activities of young workers. Physical education, the practice of sports and training will increase the physical resistance of young workers and allow them to cope more easily with their work. Manual work will be better tolerated if the young worker is "in good shape" and his capacity to adapt himself to efforts, environmental changes and aggressive factors of all kinds has been properly developed. From this aspect, it can be considered that certain judiciously chosen physical activities can help to increase resistance to work-induced fatigue.

Naturally, there is a limit to what should be attempted, and in adolescents particular care must be taken to avoid reaching a condition of physical overstrain and exhaustion, recovery from which can be protracted.

The acquisition of good sensory, motor and psychomotor capacities can be facilitated by appropriate physical activities. Dexterity, speed and precision of movement and the sense of balance can be improved, while behaviour and nervous equilibrium will also benefit. These improvements will not fail to have a positive effect on efficiency and output at work, as well as on occupational safety.

If the young worker is in a sedentary occupation, physical education and sport take on a vital importance. He needs to spend his energy and to have a normal physical activity for his age. Fatigue due to long periods in a sitting posture, or nervous fatigue due to intellectual work, can be effectively countered by participation in sports, which often constitutes an effective therapy against fatigue induced by a rapid pace of work accompanied by substantial nervous strain.

Well-trained subjects in full possession of their motor, psychomotor and sensory faculties and whose reflexes are rapid and accurate will be less prone than others to occupational accidents. A person "in good physical form" who adapts his efforts to the purpose aimed at, and is versed in the appropriate movements and techniques will behave more safely and will find himself less frequently in dangerous situations; thus he will be forearmed against accidents.

Importance can also attach to the teaching of various ways of saving oneself when a fall becomes inevitable, such as the techniques taught in connection with balancing exercises and the practice of judo. If one bears in mind the fact that falls account for a considerable proportion of all occupational accidents, it will be understood that such techniques can directly help to minimise the seriousness of such accidents, or to avoid an incident turning into an accident; this is particularly true of commuting accidents, including slipping, stumbling against an obstacle, or falling from a bicycle or motor cycle, etc.

Gymnastics and sports can also be used to prevent the occurrence of certain work-related deformities or affections. Particular stress must be placed on the importance of developing strong abdominal and back muscles in young persons. This plays an essential part in preventing bad work postures, and their evolution towards such anomalies as hyperlordosis, kyphosis and scoliosis. Strong back muscles will also help to prevent the appearance of acute or chronic lumbago. Strong abdominal muscles certainly help to minimise gynaecological disorders in adolescent girls leaning over machines and exposed to vibration, while in young men it will help to prevent stomach disorders in earth-moving equipment operators and truck drivers, or hernia in building construction workers, for example.

While physical education should be part and parcel of the general education of adolescents, vocationally-oriented physical education should be no less a part of the vocational training of young workers. The latter should be trained in the movements and techniques that they will use, with a view to achieving maximum efficiency with minimum effort. An obvious instance, for example, is training in safe and physiologically appropriate methods of handling loads. The accent should be placed on development of the physical qualities which will be most needed in the present or future occupation. This can only have beneficial effects on productivity, safety and adaptation to work.

Special reference should be made to gymnastics effected repeatedly during breaks in the working day, at the most appropriate moments, i.e. moments when fatigue becomes noticeable. Such gymnastics are physically designed to combat sedentariness, decrease nervous tension and create a psychological diversion. They consist of simple exercises demanding little effort, the aim of which is to relax the worker and not to make him feel more tired still, and which are adapted to the work effected and the prevailing working conditions; they are performed during a break of about 10 minutes. Such "active breaks" may sometimes include games, table tennis, swimming, etc. This method of combating fatigue in industrial workers is important from several points of view (welfare of workers, improvement of quality, output and safety of work). Relaxation techniques can also be resorted to. Such work breaks provide an opportunity, as it were, of "recharging the body's batteries".

Finally, physical activities are also important for handicapped workers, as a means of reducing the existing deficit, combating any tendency to general weakness and providing psychological reassurance, by showing them what they are capable of doing. The rehabilitation of functions and motor capacities and the general and vocational retraining of handicapped persons or occupational accident victims are vast problems with which industrial physicians cannot fail to be concerned.

The occupational health specialist must not only be aware of the needs of young workers, but must make them known to others, stressing the fact that adolescents must have a physically healthy activity in agreeable surroundings, and not just a job - sometimes physiologically ill-suited - in the polluted air of a plant. He must promote action to give all young workers a chance of participating in physical education or sports. He must emphasise the advantages of planning physical activities in the light of occupational needs, and must give appropriate guidance based on his knowledge of adolescents and of their workplaces. At periodic examinations, he will be able to advise young workers to take up such and such a sport, in the light of their characteristics and of any deficiencies requiring compensation or aptitudes the acquisition or development of which would improve their fitness for their job. It would even be very useful if the works doctor could examine the young workers under his care as to their fitness for sport and co-operate with physical education instructors or supervisors, for example by establishing a common file (mentioning contra-indications for certain exercises or particular indications for others). Close co-operation should also be set up between the works medical service and persons responsible for teaching load handling and transport techniques and ergonomic working methods. Young workers should be acquainted as soon as possible - preferably already during their vocational training - with these methods of work.

Sleep During Adolescence

The average duration of sleep decreases progressively from early childhood to adulthood. The principal activities of infants in arms are sleeping and eating; young children need long periods of sleep, and in a French survey it was found that during the prepuberal period, the average duration of sleep is about 9 1/2 hours, after which it decreases progressively to 7 1/2 hours, the average period in adulthood, which is reached towards the end of adolescence. The amount of sleep which young people should get varies with age and can be fixed approximately at 10 1/2 hours at age 10-12 years, 9 1/2 hours at age 13-14 years, 9 hours at age 16 years and 8 1/2 hours at age 17 years, which still exceeds adult requirements (according to Helbrügge).

The sleep requirements of adolescents are thus greater than those of adults. The "physiological work" of growth and the energy expenditure resulting from physical, mental and affective activity must be compensated, and in this respect sleep plays an essential restorative role. In these circumstances, there is no doubt that night work, with the disturbances of sleeping habits and additional fatigue that it involves, is especially bad for adolescents, and indeed, the international standards provide for the banning of night work for children and adolescents.

Sleep not only has a restorative function but is extremely important for good mental health and nervous and affective balance. Dreaming appears to be accompanied by processes of selection, classification, comparison and evaluation of information received during the day; it will be understood that in an adolescent, who is deeply involved in acquiring knowledge and affective values, this activity is especially important and must play a notable part in processes of mental development and maturation.

It must be emphasised that individual variations are greater than age-related changes in average duration of sleep. Whereas all young children sleep a great deal and variations in the length of sleep during the prepuberal period are comparatively small, in the course of adolescence more marked individual variations appear, while among adults, one finds some persons who normally sleep 5 hours nightly, many who normally sleep between 7 and 8 hours, and some who normally sleep up to 9 1/2 hours every night.

The duration of sleep in adolescents will vary, depending on the amount of physical or mental energy expended during the day. Various other factors, however, such as family habits, also play a part. The time of the evening meal affects the time of going to bed (about an hour and a half to two hours later), and when adolescents stay up late, the hours of sleep lost are only occasionally made up next morning. Lengthy spells spent watching television can, if they become a habit, appreciably reduce the amount of sleep obtained by some young persons.

It may also happen that a young person sleeps normally during some periods, while at other times, he may sleep so much that his family may begin to worry about it. This kind of variation, however, appears to be fairly frequent during adolescence. In some young persons, abnormally long sleep may be related to an instinctive desire to evade the difficulties of life which are beginning to make themselves felt, or to extreme lassitude stemming from boredom or from lack of interest in the world around them. It may also be a defensive reaction against the over-aggressive influences of the noisy, hectic and unsettled lives that they tend to lead.

Many young persons have difficulty in getting up in the morning, and this is partly due to the late hours that they keep, although other factors also appear to play a part, such as environmental noise at night time (particularly road or air traffic) and possibly air pollution.

Some adolescents tend to cut down their sleeping hours to excess, in some cases in order to study, in others in order to participate in as many cultural, physical and social activities as possible and to take the utmost advantage of this new life that is opening out before them and of their youth. This reduction in sleeping hours and the consistent tiredness which frequently results often entail numerous disadvantages.

Young persons seem to find it easier to recover well after a tiring day, especially a physically tiring one, than do their seniors in whom, on the contrary, tiredness tends to accumulate. It seems that occasional very late nights are much more easily compensated towards the end of adolescence and in early adulthood, at the age of 20 to 25 years, than after the age of 30; but here again, individual variations greatly exceed the average tendencies that could be pointed to.

The works doctor will have occasion to observe at the workplace the unfavourable consequences of lack of sleep; he may meet young workers in whom an excessive need of sleep is a sign of lack of adaptation. In such cases, he must provide the necessary health counselling.

Housing Problems

Family housing. At home, growing children need increasing space for play, and their elders must be able to study or read without being disturbed by their noise and games. Separate bedrooms must be provided for boys and girls, at least, and possibly for older and younger children as well. This will depend on the size of the family, but in any case the children will need more room. It is natural that an adolescent boy or girl should want their own room; indeed, adolescents will find it easier to fit into the community if they have a retreat of their own in which they can be alone to concentrate on matters of concern to themselves.

The scarcity of housing, its poor quality, inadequacy and unsanitary character are undoubtedly factors which, where they exist, adversely affect the health of the population, particularly of children and adolescents. Deficient housing brings a higher morbidity rate in its train, and also has a very negative effect not only on physical but also on mental development, on account of the feelings of inferiority, degradation and hopelessness that it engenders, as well as of the promiscuousness of slum life. It is also obvious that defective housing conditions make adequate personal hygiene more difficult.

ILO Recommendation No. 115 concerning workers' housing, adopted in 1961, makes suggestions regarding the implementation of a housing programme, and provides in particular for the establishment by the competent authority of minimum housing standards in the light of local conditions. This is a matter of concern to numerous government authorities and international organisations; great efforts are being made in this respect and many surveys have been effected, but in a number of countries, the housing problem remains serious, on account in particular of the rapid population growth, which overtaxes the facilities for building.

Collective accommodation. In some circumstances, for example when the home is too far away from the workplace or place of apprenticeship, young workers must live in collective accommodation. The quality of such accommodation and the existence of a healthy atmosphere there are of vital importance for the well-being and protection of the young persons inhabiting it.

The international standards provide that such housing should comprise separate accommodation for men and for women, that there should be adequate lighting, ventilation and heating facilities, an adequate supply of safe water and adequate drainage, washing facilities and sanitary conveniences. In or near the accommodation, there should be common dining rooms or canteens, rest and recreation rooms, as well as health facilities.

The main responsibility of the works doctor, when workers are housed by the undertaking employing them, is to ensure that health conditions in the accommodation are satisfactory and that the relevant standards are properly applied in fact. This is of particular importance for young workers.

Spare-Time Activities of Adolescents

In addition to working hours, mealtimes, periods of rest and recovery from fatigue accumulated during the day or the week, and sleep, a person normally has some spare time in which he may do what he chooses.

This spare time may not exist when hours of work are excessive or the work itself is too hard or too all-absorbing; in such cases, all the interval between working periods will be devoted to recovering some accumulated fatigue, as occurs in conditions of "sweated labour", when work is interrupted only to eat and sleep, and resumed again on waking.

Leisure time may also be non-existent because the activities of the person concerned are so numerous, because he has family or personal problems, or simply because, lacking any personal interests and not knowing what to do with his time, he tries, perhaps unconsciously, to eliminate these "idle hours" which are a source of boredom. There are also people who never have any spare time because they do not know how to make it.

Spare time judiciously used for relaxation and recreation has an extremely beneficial effect on every person and on his work. The time spent thus by those who know "when to stop" or "when to change their ideas" is far from wasted; such persons work more productively and more efficiently, while becoming less tired and recovering more easily from the stresses and strains of work. Leisure time may be regarded as a necessary complement of work, which enables the worker to recover his mental and physical balance. It is especially important that adolescents should be able to enjoy sufficient leisure time, since they cannot find in their work everything that they require for a harmonious physical, mental, intellectual and affective development, or for the establishment of normal social relations; it is in their leisure time that all the activities which will help to bring them to adulthood and facilitate their maturation must take place. But young persons not only need spare time, they also need some guidance in making the best use of it, as well as the opportunity to do so.

A part of it, obviously, can be taken up with physical education and sports, and reference has already been made to the various aptitudes that these will enable them to perfect, while many sports, in addition, entail a beneficial contact with nature. But it is in their leisure time that they will also be able to assuage their thirst for knowledge of things and people, to indulge in intellectual activities (reading, science, visits to museums, architecture, archeology, etc.), to participate in cultural and artistic activities and to become acquainted with the world of ideas in general. On a more practical level, pursuits such as hobbies or gardening, as well as hiking and excursions, are excellent recreations and a source of personal enrichment; travel, in particular, helps young persons to broaden their ideas.

ILO Recommendation No. 21 of 1924 concerning the utilisation of workers' spare time considers inter alia that during their spare time, workers have the opportunity of developing freely, according to their individual tastes, their physical, intellectual and moral powers, and that such development is of great value from the point of view of the progress of civilisation; a well-directed use of this spare time, by affording to the worker the means for pursuing more varied interests, and by securing relaxation from the strain placed upon him by his work, may even increase the productive capacity of the worker and his output, and may thus help to obtain a maximum of efficiency from the working day. While life outside the factory should be completely free and independent, activities should be made available to workers affording them opportunities for the free exercise of their personal tastes, while encouraging the harmonious development of their faculties. Activities which have the object of improving the workers' domestic economy and family life (gardens, allotments, poultry-keeping, etc.) are regarded by this instrument as particularly recommendable.

Although these questions are perhaps not of direct concern to the occupational health specialist, he should be aware of their importance. With his knowledge of what takes place during the working day, he is particularly well placed to understand the needs of young workers which should be met, and will be able to give advice on the kind of activities which are most likely to enable each individual worker to achieve a well-balanced existence and personality. In making known these needs and in making appropriate recommendations, he will stimulate action aimed at giving young workers not only an opportunity of relaxation and recreation, but also an opportunity of personal enrichment from every point of view.

CHAPTER II

HEALTH PROBLEMS OF ADOLESCENTS

General Considerations

At pre-employment medical examinations, the industrial physician will find himself face to face with young persons who may be adolescents straight from school, who have never had a job before, or young workers who are changing jobs for various reasons, possibly including health reasons. He will have to decide whether or not these young persons are medically fit to be employed by the undertaking in the job concerned.

If his findings are positive, this pre-employment examination will become the starting point of a new responsibility involving continuing medical supervision; it will serve, as it were, as a point of reference for subsequent examinations.

At the succeeding periodic examinations, the physician will be able to follow the evolving adolescence and developing abilities of these young workers. He must do so, not as a casual spectator, but as a conscientious witness, always ready to help the young person concerned to develop towards adulthood. He must be on the lookout for any anomalies in this development, for signs of lack of adaptation to the job, as well as for any signs of ill health, job-related or otherwise, and take appropriate remedial action.

The physiological and pathological peculiarities of adolescence will be considered in this chapter, taking as a basis the conduct of the medical examinations by which the works doctor will evaluate the state of health and fitness for work of the young worker.

From the moment the adolescent enters the consulting-room, certain observations can already be made (extent of growth as regards height and weight, posture, muscularity, etc.).

The occupational health specialist will ask himself whether the subject is sturdy enough to perform the work concerned, to what extent he is capable of physical effort, and what is his capacity of adaptation to effort.

He will have to evaluate the perceptive systems (sensory organs, proprioceptive sensitivity) and the characteristics of perception and motoricity at the level of the central nervous system, being quick to discern disturbing factors (fatigue, mental problems or disorders, physical affections) which may have an adverse effect on these functions and on the sequence of perception, integration and movement.

The physician will also look for any anomalies or affections which might predispose the subject to certain occupational risks (such as dermatitis), suggest the existence of an occupational disease in its early stages, or make the work concerned more arduous (such as digestive disorders or postural anomalies). Medical examinations of adolescents must also be conducted as systematic check-ups, designed to ensure that the subject remains in the best possible health (pin-pointing of anomalies or disorders of the systems already referred to, as well as of the digestive system, teeth, genito-urinary tract, endocrine system, etc.).

General Appearance and Condition

As soon as the young worker has entered his office, the doctor can begin collecting fundamental information about him.

Among the questions that the doctor should keep present in his mind, which can be quickly answered, the following in particular are of interest:

- How does the subject walk? Is his gait confident or hesitating (possible cerebellar disorders, central nervous system involvement, vertigo, fatigue, etc.)? Do the feet contact the ground firmly, or does he stumble (medullary involvement, superficial peroneal paralysis, etc.)? Does he limp? Is the limp unilateral or bilateral? What is the reason (shortening of lower limb, club foot, disorder of the spine, hip, knee, instep or foot, congenital dislocation of the hip, etc.)? Does he appear to have flat feet? Is his gait clearly indicative of great fatigue or an obvious lack of assurance?

- How about his posture, does he stand up straight or does he lean on one leg? What is his general outline like (hyperlordosis, kyphosis, scoliosis, etc.)? Is there any tremor of the extremities?

- How does he undress? Does he lose his balance? Are his movements co-ordinated? Does he discard his clothes carelessly or is he tidy? Is there any shortness of breath?

- What is his general appearance? Does he look his age, or does he look older or younger? Could any discrepancy be related to an abnormal development of puberty? How about his build, is he tall and thin, or small and fat, how about growth of hair? Does he seem well-nourished? What is the state of muscular development? Is there any lack of symmetry?

- What is the appearance of the skin? Does his colouring appear healthy, or does it suggest some underlying affection (tuberculosis, heart disease, anaemia, liver complaint)? Is there any dermatitis or acne, and if so in what areas? Is there any cyanosis? Do the hands perspire profusely?

- What does the examination of the chest, abdomen, etc. suggest?

While questioning the subject on his medical history, also, the doctor, in addition to noting data for the record, will have an opportunity of observing certain other points of interest:

- What is his manner in replying? Does he express himself clearly and logically? Does he seem reticent or communicative? Are his replies aggressively framed? Does his behaviour suggest that in his view, the medical examination is an irksome duty which he is anxious to be rid of, or does it suggest, on the contrary, that he is concerned about his state of health?

The over-all evaluation of "normal" or "pathological" appearance, and, for that matter, the (highly subjective) evaluation of "physical resistance" and "sturdiness" play a considerable role in the prognosis to be made by the works doctor: will a particular subject fit into a particular job without harm to his health? It is not possible to found such assessments on objective criteria, and their worth will depend on the doctor's experience of young persons and their behaviour at workplaces, as well as of the workplaces themselves.

Osteo-Articular and Ligamentary System

The skeleton undergoes continuous development from birth until adulthood; this development is characterised by the coexistence of growth and maturation processes. Maturing takes place in two stages, namely, calcification of the epiphyseal centres, which terminates at the beginning of puberty, and closing of the epiphyseal lines, which occurs in a certain chronological order after puberty, and the end of which marks the end of growth. The study of bone age is a good criterion of physiological maturing in adolescents. Generally speaking, it may be said that ossification is largely completed at 16 years of age in girls and 18 years of age in boys. Subsequently, some traces of the former connective cartilage tend to be found (towards 19 to 20 years of age), and some centres of ossification subsist at the level of the spinal column.

The presence of epiphyseal lines in young growing adolescents implies that certain injuries may be accompanied by epiphyseal fractures which pose particular problems of diagnosis and treatment, which must be performed with special care, in view of the possible complications. In adolescents, the metabolism of the bony tissues is still very active, and fractures generally heal very well and quickly.

During the prepuberal period in particular, during which the skeleton undergoes rapid growth, but also at the start of adolescence, "growing pains" may be complained of, often in the area of the joints, especially the knees. Various explanations of these pains have been suggested, none of which is very satisfactory, and in the presence of such symptoms, the doctor should ask himself whether they may not be due to a "growth osteochondritis" corresponding to dystrophic disorders in the growing bone and cartilage tissues.

This kind of osteochondritis is often triggered by a severe muscular effort or exaggerated muscular activity; an example is the "anterior tibial apophysitis" which may occur in young boys following severe over-extension of the knee. Many different varieties of epiphysitis and apophysitis, depending on the location, have been described.

A clear distinction must be made between, on the one hand, "growing pains", which are harmless, or "growth osteochondritis" which responds well to rest and heals without complications, and, on the other hand, two affections which are serious and differ greatly from those just mentioned, not in their initial symptoms, but on account of their evolution and of the deformities that they may entail; these are separation of the upper femoral epiphysis (coxa vara of adolescence) and juvenile kyphosis (Scheuermann's disease).

Adolescent coxa vara is due to dystropic disorders of the bone and cartilage tissues of the upper femoral metaphysis; it may lead to a deformation and displacement of the femoral neck, with possible concomitant arthrosis of the hip-thigh (coxofemoral) joint. This disease is noted more particularly in boys. Scheuermann's disease is an epiphysitis of the vertebral bodies, and also stems from an osteo-cartilaginous dystrophy of the areas of growth; it may lead to a painful kyphosis. In both cases, there will be a history - either as the initial cause or as an aetiological factor - of local overstress resulting from hard labour (such as the carrying of heavy loads) or again, in the case of Scheuermann's disease, of overstress incurred by spending too much time in a sitting or standing position. It follows that these diseases are of direct concern to the occupational health specialist, all the more so since, as their onset is insidious, it is he who will be in possession of the first evidence of their possible existence.

Signs of rickets may be found during the examination. Although this is intrinsically a disease of childhood, the physician may sometimes have occasion to wonder whether the lesions noted are effectively stabilised and whether the progress of the disease has been halted; does the subject's diet include a sufficient and balanced component of phosphates and calcium, and is the calorie and protein intake adequate? Supplementary examinations may be required.

Tuberculosis of the bones is more frequent in children and young adolescents than in adults, perhaps because the growing bony tissues, which are highly vascular, offer a favourable environment to the tuberculosis bacillus. Localisation frequently occurs in the area of the spinal column; tuberculosis of the joints is also found. The occupational health specialist must keep these possibilities in mind, for the evolution is frequently insidious, and the establishment of a clear diagnosis takes some time.

In adolescents, the joints are supple and their ligaments are strong. Sprains usually heal very well. The joint cartilages are well supplied and are not prone to degenerative processes, although it should be noted that degenerative lesions of the intervertebral discs and radiological signs of osteophytosis may already appear at the age of 20 or 25 years. The presence of back or lumbar pains may also suggest the possible late appearance of a congenital anomaly of the spinal column.

In adolescence, also, Still's disease (a specific form of acute rheumatoid arthritis occurring mainly in childhood) and initial signs of ankylosing spondylarthritis (towards the end of adolescence) may be found.

Muscular effort is transmitted to the bony system via the connective structure arranged round the muscle fibres, and via the tendons; the joints and ligaments also play their part in transforming muscular contractions into movement. All these elements must be in good condition in order to perform efficiently the movements required in life and work. Vice versa, also, the good condition and resistance of the connective tissues, as well as calcification, to some extent, and the nutrition and metabolism of the cartilages, depend on the movements performed, and the mechanical actions to which they are habitually subjected. Thus it is understandable that spontaneous physical activity - which is generally very marked in young persons - is a particularly important factor at a time when these tissues are still in full process of growth and development.

Persons with infirmities of the locomotor system which are not of neurological origin, that is to say, persons who are handicapped posturally or in their motor and locomotor capacities, not as a result of a lesion of the nervous system, but in consequence of some congenital malformation (club foot, congenital dislocation of the hip, etc.), an injury (amputation, pseudarthrosis, etc.), or a disease of the bones or joints (resulting from tuberculosis of the bone, osteomyelitis, etc.), pose special problems of vocational training and job adaptation which directly concern the occupational health specialist. He will also have to go into the choice of any artificial limbs or appliances that may be required, and be ready with information and advice to ensure that the choice is made not only in the light of the needs of daily life, but also and perhaps especially bearing in mind occupational requirements and possibilities. In addition, special medical supervision will be required to evaluate and follow up the degree of job adaptation, noting the beneficial or harmful effects of the work involved on the handicapped person.

Locomotor System and Postural Anomalies, Back Pains

The diagnosis of locomotor system and postural ancmalies is particularly important in young workers for several reasons. These subjects are at a stage in which any postural anomalies are, as a rule, not yet irreversible. There are as yet no definitive lesions or substantial pains. If no remedial action is taken, however, the defective postures will become established. The changes in the bony structure which still take place during adolescence entail the dual possibility of speedy deterioration on the one hand and of relatively easy correction on the other hand. These circumstances must be taken into account, since by taking early action one can, for example, prevent the appearance of many kinds of back pains resulting from postural anomalies. The importance of this matter is underlined by the frequency of back pains and their influence on sickness-induced absenteeism, and on fitness for work (the fear of an acute attack or the persistence of chronic pain may finally lead to real invalidity).

Locomotor system and postural anomalies are particularly frequent in adolescence. Growth in height and weight and muscular development are not always balanced, by any means, and this imbalance favours the appearance of spinal deviations, for example. In addition, the great flexibility of the ligaments that still exists may tend to favour defective postures where muscular support is insufficient. During pre-employment and periodic examinations, therefore, the doctor must be on the lookout for locomotor system anomalies, and make a record of any that may be fcund. Such examinations may be made very quickly by those accustomed to do doing them, and need no special equipment. At periodic examinations, the evolution of anomalies previously noted should be checked, and if any new anomalies are found, the possibility of their being due to conditions at the workplace (defective working posture) should be investigated.

Adolescents are prone to flat feet, but this condition is generally painless. It may be related either to some slackness in the ligaments or to a weakness of the flexor muscles. A good general preventive measure is to teach young persons to carry their weight not on the heels, but on the soles of the feet. One can also more or less identify a particular variety of flat foot with concomitant pain, "adolescent's tarsalgia", which is noted principally in boys and in which there is a progressive flattening of the arch, with osteo-articular lesions of the tarsus which finally become irreversible. Substantial job placement problems may occur in the case of young workers with fallen arches, for whom long periods spent standing are unbearably painful or may aggravate existing lesions, while the wearing of protective footwear may also pose problems. In young girls, the wearing of high-heeled shoes brings a number of hazards in its train, such as stumbling and falling, increased risk of lumbar lordosis and back pain, and a tendency to wear shoes which are too small or too narrow can cause local fatigue and possibly lead to deformed feet.

In examining the spinal column, it may be found that certain physiological curves are unduly marked, as in lumbar hyperlordosis and dorsal kyphosis. More rarely, on the contrary, a decrease or even an inversion of the normal curvature may be found. Lateral deviations of the vertebral column (scoliosis) may be noted. It is important to distinguish between lateral deviations which are corrected by bending forward and those which are not; the former are scoliotic postures, while the latter are cases of scoliosis properly speaking.

Lumbar hyperlordosis is extremely frequent in adolescents, particularly in young girls, in whom the normal curvature is already more marked than in boys. The cause is in most cases a muscular insufficiency of the abdominal wall, which allows the pelvis to tilt forward, with a compensatory hyperlordosis; other factors also enter into consideration, in particular insufficiency of the dorsal extensor muscles, relaxed ligaments and in some cases, certain kinds of work which are detrimental, such as the carrying of heavy loads.

Dorsal kyphosis is frequently found in conjunction with lumbar hyperlordosis. The curvature of the dorsal column is accentuated in order to compensate the excessive curvature of the lumbar column and thus maintain posture; other factors may contribute to this tendency, in particular lack of tone or insufficiency of the shoulder girdle muscles, the shoulder blade fixation muscles and the dorsal extensor muscles, or the development of the chest muscles which tend to pull the shoulders forward. The result can be a very ungraceful, hunched-up silhouette, with the head and neck projecting forward, the shoulder blades detached and wide apart, and a flattened chest. This is an asthenic attitude, in which the subject does not use his muscles to maintain posture, but rests on his ligaments. Although such deformations may be corrected for long periods by voluntary muscular effort, ultimately the anomalies of spinal posture, as a result of the abnormal pressure to which they give rise at the level of the vertebrae, will lead to deformations which will become irreversible and will favour the occurrence of vertebral arthrosis due to overstress. It must always be remembered that lumbar hyperlordosis and dorsal kyphosis may be symptomatic of a congenital vertebral anomaly or of a developing affection of the spinal column.

Some adolescents adopt another kind of asthenic position. They "rest" on the anterior hip ligaments, which are very strong, in doing so causing the pelvis to tilt; it is as if the subject was "sitting in his pelvis". This tilting causes a lumbar kyphosis, with concomitant dorsal kyphosis.

At workplaces, care must be taken to ensure that working postures and conditions do not encourage the tendency, already widespread, to such deformation. For example, the height of the worktable or bench, and the height and design of seating, must be suited to the height and build of the adolescent. But adolescents must also be taught to take up correct postures; effective training is indispensable if the other measures taken are to retain their full value. In some cases, vocational counselling or redirection may even become necessary. It must be remembered that in some occupations, kyphotic postures are very frequent, as in high-precision manual work performed sitting or standing (watchmaking, assembly by soldering in the electrical and electronics industry, etc.), sculpturing, cabinet-making, shoemaking and repairing, lathe work, and load carrying using the shoulders; long periods spent standing up and bending forward are particularly harmful.

The scoliotic posture is characterised by the appearance of lateral deviations of the spinal column, which may be corrected by various manoeuvres, principally by forward flexion; they do not indicate a malformation of the spine bones. The possibility should always be borne in mind of a difference in the length of the lower limbs, producing a lateral tilt of the pelvis and compensatory scoliotic curvatures at the level of the spinal column; in this case, the scoliotic posture is corrected when the subject is seated. But it may have originated at school, due to the adoption of an unsuitable position for writing or sitting; this problem has been treated in many studies of health in schools. Unfavourable positions may also be taken up during work, in which case redesign of the workplace and postural training may be necessary. Another important factor which always crops up is the existence of muscular insufficiency or weakness leading to the adoption of unsuitable positions; asymmetrical muscular development resulting from the work performed may also play a part. Here, symmetrical reinforcement of the muscular system by appropriate gymnastics is of great importance.

"Idiopathic scoliosis" is a specific disease of adolescence. It begins somewhat before or around the time of puberty and becomes consistently worse until the end of growth. It is more frequent in girls than in boys, and is characterised by vertebral lesions, and principally by the existence of a rotation of the vertebrae with respect to one another. The real cause remains unknown, although various theories have been advanced. The vertebral anomalies are responsible for the appearance of primitive curvatures, the place and form of which are variable, and which in their turn induce an adaptation of the spinal posture and the appearance of compensatory curves. This posture may be balanced or asymmetrical, as may be checked by suspending a plumb-line from the nape of the neck; if it falls to the right or the left of the gluteal fold, this may point to an assymetry and to persistently developing scoliosis. The diagnosis may be confirmed radiographically and in particular radioscopically (study of mobility of spinal column). The taking up of an occupation may hasten the development of an idiopathic scoliosis and involves the appearance of pain induced by muscular or ligamentary overstress, whereas this scoliosis normally takes a painless course. Thus, the employment of such persons is a delicate problem; it is necessary to avoid heavy work and work which leads to fatigue of specific groups of muscles, as in production line work; long periods spent sitting down are usually not well withstood and it is preferable for young persons with scoliosis who are able to work to be given light and varied jobs, with frequent changes of posture and movement.

Scolioses can also be described which are due to lesions of the spinal column, but which do not develop in the same way as "idiopathic scoliosis". These include scolioses rsulting from rickets or from some congenital or acquired vertebral anomaly, which may or may not be evolutive. In the long term, scoliotic vertebral lesions may be observed, resulting from disorders of vertebral posture, from a fixed scoliotic attitude, from paralyses of the dorsal extensor muscles following infantile paralysis, or from attraction of the spinal column by a fibrothorax or by substantial pleural symphyses.

Lumbar or dorsal pains, whether acute or dull, transitory, repeated or chronic, are one of the conditions most frequently complained of by adults and old persons, particularly women. Although such symptoms are frequently trivial and related to incipient influenza, to symptoms of infection or to a state of unusual fatigue, and although their diagnosis is sometimes easy (the result of strain or effort), the fact remains that in many cases the diagnosis of back pains is a complex matter, as there are so many possible causes (muscles, ligaments, bone injury, prolapsed disc; dorsal, urinary, digestive or gynaecological; accidental, infectious or inflammatory origin, etc.).

Back pains are not a rare occurrence in adolescents, especially in girls. Although all the possible diagnoses must naturally be considered and whatever additional examinations may seem advisable must be performed, many cases may turn out to be trivial or due to some orthopaedic anomaly (particularly different lengths of lower limbs), badly performed muscular efforts, or to an anomaly of spinal posture (hyperlordosis, scoliotic attitude), itself often the result of muscular insufficiency or simply of bad postural habits. The occurrence or reoccurrence of such back pains may be prevented by quite simple means. The action which should be taken - and which should be familiar to all concerned - includes in particular:

- general health measures, at home (adoption of a correct posture when sitting or standing, appropriate bedding) and at the workplace (layout of workstation and conformity of seating with biometric measurements, to permit and encourage the adoption of physiologically correct postures);

- regular performance of appropriate exercises (involving stretching, straightening and bending of the spinal column, etc.), with suitable gymnastics during work breaks at the workplace;

- performance of corrective or formative gymnastics to strengthen the muscles of the back and abdomen, or participation in sports known to have similar effects;

- the use at the workplace and away from it of appropriate methods of handling, lifting and carrying loads.

A physician who pays due heed to anomalies of the locomotor system and of spinal posture can assist a great many young persons and effectively help to minimise the incidence of affections which are becoming one of the major problems of medical services, occupational and otherwise.

Muscular System and Muscularity

In adolescence, the muscular system undergoes substantial development, which is influenced not only by constitutional factors, but also in a great degree by nutrition, physical exercise and training. This development is marked by wide variations between individuals, as well as by increasing differentiation between the sexes.

Muscular development is important for adolescents, not only from the physical point of view, for the purpose of engaging in the activities which are normal at their age in private and occupational life, but also for psychological reasons and even from the standpoint of social adaptation, so great is the importance that adolescents attach to their muscular capacity.

Muscular strength increases with age from childhood, reaching a maximum between the ages of 20 and 30, depending on the muscle groups considered. In boys, muscular strength measured dynamometrically increases by over 100 per cent between the ages of 12 and 18 years. The increase is more rapid in boys than in girls; from the onset of puberty, the difference becomes progressively more marked, until at about 18 years of age, the muscular strength of girls is on average only about 60 to 65 per cent of that of boys. The lower limbs are the first to gain in strength, while the prehensile strength of the hand is practically doubled in boys between the ages of 14 and 18, reaching a maximum at about 20; in the case of the back muscles, maximum strength is reached a little later.

Attributing a value of 100 per cent to the muscular strength of a young male adult of 25 years of age, the development of the muscular system may be described as follows (according to Hettinger): at age 10 years, it is 40 per cent in both boys and girls, at age 14 it is 60 per cent in boys and 50 per cent in girls, and at age 18 it is 90 per cent in boys and 60 per cent in girls (in whom any subsequent increase will not exceed a few per cent). These data are of assistance in studying the adaptation of work to young workers, or the reasons why an apparently normal job may prove excessively tiring. It is generally considered that in order to preclude the appearance of symptoms of fatigue, repeated efforts exceeding 15 per cent of the maximum capacity should be avoided; thus, in order to operate a lever by flexion of the forearm and upper arm (maximum capacity about 30 kg in adults), an adult should not have to exert a force of more than 4.5 kg, whereas in the same conditions the effort required of a 14-year old adolescent should not exceed 2.7 kg (Hettinger).

A distinction is made between static muscular strength (isometric muscular contraction) and dynamic muscular strength (exerted during movement). A parallel increase in both types of muscular strength occurs with increasing age; static muscular strength may be measured dynamometrically, and it may be of interest to follow its development during growth. Static muscular contraction (with blocking of the blood flow in the muscle) is much less favourable physiologically than the performance of movements, and adolescents have even more difficulty in withstanding it than do adults.

The extent of the muscular strength that can be exerted depends on a whole series of factors; in particular, it is a function of the mechanisms producing muscular tension, of the location of fulcra and points of resistance, of the "leverage" of the muscle system used, of the anaerobic energy capacity (in isometric muscular contraction, the circulation is blocked) and of the glycogen storage capacity and aerobic capacity (during movements), as well as of vascularisation, co-ordination of agonist and antagonist muscle action and in particular of the activity occurring simultaneously in these two muscle groups during a movement. Wide constitutional and individual variations are encountered. Some of these factors may be modified by training, which helps in particular to improve their co-ordination in the exertion of muscular strength.

The repetition of muscular effort at a given rate gives rise to local muscular fatigue which at a certain point hinders the correct performance of the exercise; this provides a yardstick of muscular endurance. The latter can be improved in particular by repeating a series of efforts daily over a certain period. Such training is more effective in adolescents than in adults. The period between 12 and 15 years of age is especially propitious to the development of good muscular resistance, which continues to be displayed for several weeks after the end of training. This suggests the occurrence of lasting modifications - involving vascular changes with an increase in the number of capillaries - within the muscle group involved. Plethysmographic measurement has shown that the increase in blood flow in trained muscles during exercise is much more marked in young subjects than in older persons.

In the field of muscular pathology, reference should be made to "myasthenia gravis" which may develop during adolescence, or become more evident then in latent cases. Muscular dystrophy, such as Erb's disease, may be encountered, and it is reported that "myositis ossificans progressiva" may begin at this stage of life. Most frequently observed, however, are muscular bruises and minor cases of torn muscles sustained as a result of intense physical and sporting activity; these usually heal very well, but may tend to recur.

Cardiovascular System

The Heart

Adolescence is a time of growth and great physical activity. The heart also develops, its volume being almost doubled between the ages of 11 and 16 years. During this stage of life there is great cardiac activity; the faculty of adaptation to effort is considerable and the maximum heart rate reached during effort is substantially higher than in adults. On the other hand, since the systolic output is lower than in adults, a greater increase in heart rate is required by adolescents in order to achieve the same increase in blood flow and the same amount of oxygen; thus work which entails an increase in pulse rate of three beats a minute in adults will entail an increase of seven to eight beats a minute in a 14-year old adolescent (Hettinger). It follows that in a young worker, the heart is put to a greater contribution than in his seniors.

The mean pulse rate at rest increases regularly with increasing age in both boys and girls (Nelson). In adolescents, the heart reacts rapidly and markedly to physiological or psychological stimulations; these reactions may be said to be correlated, at least to some extent, with the hyperexcitability and lability of the sympathetic nervous system which are so often encountered in adolescents.

The anatomical changes in the heart and thoracic cage influence the outcome of clinical examinations of the heart by palpation, percussion or auscultation. Thus a pathological increase in heart volume, for example, which appeared evident in childhood becomes much less so in adolescence owing to the enlargement of the thoracic cage; while an intensification of known heart murmurs may be noted, this does not necessarily mean that the situation has deteriorated.

A wealth of different functional murmurs may be observed at this stage of life. These can generally be heard in the second or third intercostal spaces, along the left hand side of the sternum, they are scarcely diffused and often of short duration, and occur only during a part of systole; they are liable to vary with attitude, respiration or exercise, as well as in time. Anorganic murmurs of no pathological significance are said to exist in 30 to 60 per cent of normal adolescents. It is clear, therefore, that the occupational health specialist will frequently be faced with the diagnosis of cardiac murmurs, which must be performed with particular care, since the fitness of the subject for work is at stake, and on the one hand the possibility of a heart disorder must not be disregarded, while on the other hand the adolescent and his or her parents must not be given grounds for unjustified anxiety, nor must unnecessary restrictions be imposed on fitness for employment.

A wide variety of heart disorders may be encountered in adolescents. While they include congenital diseases which have or have not yet been diagnosed, and rheumatic valvular disorders, the most frequent findings relate to functional disorders pointing to neurovegetative dystonia and symptoms of cardio-vascular erethism. The subject may complain of palpitations, precordialgia or breathlessness following effort, and considerable psychic involvement may be observed, evidenced by nervousness, hyperemotivity and, frequently, considerable anxiety; the examination may show a rapid pulse, extrasystole, possibly a raised arterial pressure and a functional murmur. It should also be mentioned that adolescents fairly frequently complain of short-lived stinging sensations or even sharp pains (very brief and sometimes very intense) in the precordial region, which are not brought on by any particular cause and are accompanied by no other symptoms; this rather trivial phenomenon has no pathological significance and has never been satisfactorily explained.

At the pre-employment examination, the physician will wish to verify whether heart function is normal or not. To conclude that the heart and cardiac functions are entirely normal and that any anomalies observed are purely neurovegetative, or on the contrary that a heart disease is present, is no simple matter, and yet in the interests of the young person concerned, it is necessary to establish a clear position, without resting content with a mere finding that, for example, "I can hear something which is not quite normal, but it isn't significant", prudently leaving room for a measure of doubt which can only be prejudicial to the adolescent. It is very important to arrive at a correct diagnosis, since over- or under-estimation may have serious practical consequences for the young person concerned. In the former case, an unjustified limitation will be placed on his activities and scope for employment, by turning him into an artifical invalid, while in the latter case the way would be open to aggravation of an unrecognised heart disorder, leading perhaps to acute attacks, whereas a quieter way of life might have resulted in a more favourable evolution.

When the signs found during the examination justify this, the examining physician should order additional examinations to confirm the diagnosis and determine the functional capacity of the myocardium. As a rule, this will be done in specialised centres. Sometimes, depending on possibilities, some additional examinations may be performed in the occupational health service such as X-ray examinations of thoracic organs (radioscopy and radiography), electrocardiographic examinations at rest and under effort, and a functional examination including in particular effort tests. When various additional examinations can be made in the medical service, this will permit an initial assessment of the anomalies noted, revealing the absence of any pathological heart condition in many young persons, and shedding more light on the nature of the affections of those who are suffering from some disorder.

Where the case is one of cardiocirculatory erethism or neurovegetative dystonia, the adolescent should in no event be given the impression that work might be harmful for his heart, for that would only make matters worse. On the contrary, he must be allowed an absolutely normal level of activity, and indeed work may have a very beneficial effect on his condition. A progressive approach may sometimes be necessary, however; thus the initial tasks should not require too rapid a pace of work or entail too substantial responsibility, and should be performed in a congenial and calm atmosphere free from too much noise and agitation, and leave room for work breaks. Apprenticeship, in which several of these conditions are fulfilled, therefore often proves beneficial.

In the presence of a clearly established heart disease, the questions of employment in particular posts and limitation of activities must be gone into in each individual case and discussed with the consulting physician. Account must be taken of the general and functional repercussions on the heart of the disorder found. Great importance attaches to psychological factors in helping the young person to accept his disability. On the occasion of periodic examinations, the occupational health specialist must arrange for a special supervision of such adolescents.

Certain types of work are contra-indicated for workers with comparatively serious heart conditions, such as heavy manual labour (miners, dockworkers, unskilled labourers), work at heights (structural assembly work, work on scaffolds), work with certain dangerous machines, operation of vehicles (cars, trucks, trains, etc., with a total ban in the case of heavy trucks and public conveyances), work performed in the standing position only (peripheric stasis), and jobs involving any responsibility for the safety of personnel (firemen, safety teams).

Arterial Pressure

Arterial pressure is on average lower in adolescents than in adults. At 13 or 14 years of age, the maximum is 11.5 cm mercury column ± 2, and the minimum is 6 cm mercury column ± 1. Systolic pressure increases throughout childhood and somewhat more quickly during adolescence, reaching adult values at 18-20 years of age.

An adolescent is in a state of growth, and a complete balance is not permanently maintained between bodily development, circulatory blood volume and the cardiocirculatory system; thus anomalies of pressure will be found fairly frequently. There may be either hypo- or hypertension, mainly systolic, moderate and transitory. Moreover the excitability of the neurovegetative system already referred to may be responsible for such phenomena or for generally irregular blood pressure.

Where high blood pressure is found, the doctor should first verify whether this is not due only to emotional reactions related to the medical examination. The importance of emotional factors must not be underestimated, especially in adolescents. Raised blood pressure may also result from anxiety or nervousness related to work or the circumstances in which it is performed; the possibility of such a cause should be investigated by appropriate questioning. But if the pressure remains high even at rest, for example when arterial pressure is measured in the morning before getting up, the cause must be found and the necessary additional examinations performed. The physician should insist on the necessity of such examinations, particularly since in some cases there is a real possibility not only of improvement but of cure (stenosis of the aorta or of a renal artery, phaeochromocytoma); but this is not always an easy matter, as hospitalisation is often necessary and frequently the adolescent is not yet suffering from any symptoms when the hypertension is found.

The following may be considered to be the limiting values if the arterial pressure is taken in satisfactory conditions and with appropriate equipment (Rutenfranz): in adolescents 14-16 years old a maximum of 13.5 or more is suspect, and a maximum of 14.5 or more is pathological; in 18 year old adolescents, a maximum of 14 or over is suspect, and a maximum of 15 or over is pathological. A diastolic pressure of 9-9.5 or more is suspect, and becomes pathological upon reaching 10-10.5 or more. In adolescence, the most significant causes of hypertension are glomerulonephritis (especially if it becomes chronic) and its complications, congenital renal anomalies (in particular polycystic kidneys), stenosis and certain other anomalies of the renal arteries, aortic coarctation (hypertension in the upper part of the body, frequently accompanied by headache and hypotension in the lower limbs, often with concomitant cramps), aortic insufficiency (raised systolic pressure but low diastolic pressure, bounding pulse and frequent fainting, particularly following effort), Cushing's syndrome and phaeochromocytoma. In most cases, however, it will be impossible to pinpoint the aetiology of the hypertension and the diagnosis will refer to "idiopathic hypertension". Problems of vocational guidance and placement arise in such cases, for the young persons must not be exposed to conditions which might induce a substantial increase in blood pressure (heavy physical work, continuous nervous tension) or to a sudden rise of pressure (effort in the thoracico-abdominal region, violent emotion).

Adolescents frequently complain of various symptoms which may be related to low or unstable blood pressure, with inadequate regulation of pressure variations during changes of posture, or hypotension after remaining standing for long periods; these include an impression of momentary blackout (particularly during changes of posture), fatigue, difficulty in concentrating, indolence or irritability, nausea, occasionally headache, or repeated feelings of faintness or fainting spells. High blood pressure has been the subject of much more research than low blood pressure, for which no satisfactory classification has been established; as a result, the terminology is not very precise, and the condition known as "orthostatic hypotension" (occurring after long periods in an upright position) may be interpreted extensively or restrictively, depending on the author.

A frequently-observed anomaly is an abnormally slow rise in physiological pressure when changing from the prone or sitting position to the upright position. Sometimes a more or less pronounced (1-2 cm mercury column) and protracted (15-30 seconds to 2-3 minutes) drop in pressure occurs, although the pulse rate quickens noticeably; there may be a simultaneous drop in maximum and minimum pressure, or a drop in maximum pressure only, with a flattening of the differential pressure and a very marked decrease in the pulse volume; this explains the impression of blackout, or the outright fainting (in marked cases) experienced during changes of posture. This sign should be investigated (it is easy to take the arterial pressure again, after first taking it in the prone or sitting position); it implies a disorder of the pressure regulation mechanism which may sometimes be related to a state of great fatigue, or to a poor physical condition with lack of activity. If an effort test is made, marked hypotension may sometimes be observed directly after the effort; this reaction is normal after a thoracico-abdominal effort, but not after a dynamic effort, when it points to a clearly inadequate adaptation to effort, a total lack of training or excessive fatigue; more rarely, it may be related to some organic disorder.

Orthostatic hypotension is particularly frequent in adolescents of 14-15 years of age. In these subjects, there is a history of headaches, palpitations and malaise, sometimes accompanied by nausea, faintness or fainting fits, after prolonged periods spent standing. The examination may show up disorders of pressure regulation; after taking the pulse and blood pressure at rest in the prone position, the subject should be made to stand up, and the arterial pressure and pulse should be taken, first after 1-2 minutes and then at regular intervals. In order to be sure that the findings are negative, the subject should be kept standing for at least 8 minutes, and preferably for 10-20 minutes. A diagnosis of orthostatic hypertension is justified if the blood pressure drops by at least 20 mm mercury and the pulse rate increases by at least 40 beats per minute (Rutenfranz).

Further if there is a history of repeated fainting fits, the possibility of hyper-reflexia of the carotid sinus must also be considered. In this case, it may be found that the faintness or fainting fits occur when the head is moved quickly or the neck is rotated and extended. This finding may be confirmed by examination, in particular by cautious manual stimulus of the carotid sinus.

The existence of disorders of pressure regulation, and especially of a tendency to fainting fits, when these are at all frequent, must be taken into account when considering job placement, in view of the hazards involved in work at heights or near dangerous machinery, etc. Work involving long periods spent standing (e.g. salesgirls) is contra-indicated for subjects with orthostatic hypotension, owing not only to the discomfort entailed, but also to the extra burden imposed on the heart, which in the long run may overload it.

Venous System

Although varicose veins in the lower limbs are encountered mainly in adults, they can also occur in adolescents, and in that event may sometimes constitute a contra-indication for certain kinds of employment (work performed standing for protracted periods or involving a risk of bruises, cuts or other injuries of the lower limbs). It should moreover be pointed out that in some cases varicose veins are listed among the grounds for unfitness noted at pre-employment medical examinations performed in conformity with the national laws and regulations on the medical examinaticn of young workers (United Kingdom).

The varicose veins may be the consequence of a congenital anomaly of the venous system, and occasionally of an arteriovenous aneurism. The additional examinations required to arrive at a diagnosis should be performed, and the subject should be advised to seek appropriate treatment from the competent medical services.

It is not unusual to find that certain young persons, towards the end of their adolescence, are already suffering from haemorrhoids and have already had acute attacks; as a rule, however, this information must be elicited by questioning. Appropriate health advice (measures against constipation) should be given, and additional examinations and medical or surgical treatment should be considered according to the case.

Cardiac and Circulatory Changes During Effort

A distinction must be made between changes in heart function and blood circulation occurring during static and during dynamic effort.

If the static effort is highly localised (as in grasping a hammer, for example), the changes will be purely local. The circulation in the statically-contracted muscles will be hampered, leading to partial anoxia and quite quickly to local fatigue. If the effort to be made or the posture to be taken up require the static contraction of large groups of muscles, the effect on the circulation may be sufficient to have general repercussions; the arterial pressure may drop and the pulse may accelerate.

In heavy work, such as the lifting or pushing of loads, etc., a thoracico-abdominal effort, consisting in a forced expiration with the glottis closed, is necessary in order to make the spinal column, thorax and abdomen rigid enough to serve as a fulcrum for muscular work performed by the upper limbs. This effort gives rise to a blockage of the venous return flow and to various haemodynamic changes. Static efforts hamper the physiological adaptation of the circulation and are undesirable; repeated thoracico-abdominal efforts may have pathological consequences, and this type of effort is particularly harmful for young persons.

In conditions of dynamic effort, on the contrary, an entirely different situation obtains. The whole of the cardio-circulatory system is brought into play in transporting a volume of oxygen and carbon dioxide which may be 10 to 20 times as great as the volume handled when at rest. There is an increased cardiac output, resulting from a simultaneous increase in heart rate and systolic output. The blood vessels of the muscles and skin become dilated, resulting in improved oxygen transport to the working muscles and a decrease in peripheral resistance, thus leading to a substantial increase in blood circulation with a reasonable increase in perfusion pressure. During the effort, a steady-state condition is reached relatively rapidly, at a level depending on the nature and extent of the effort. Cessation of effort is followed by recovery during which the heart rate decreases rapidly at first, and then more slowly; the time taken by arterial pressure to return to normal depends on the force which has been developed. Recovery is more rapid in subjects accustomed to effort than in sedentary subjects.

Occupational health specialists should investigate the cardiac and circulatory adaptability of young workers more carefully than is usually the case. An appropriate effort test should be part of the standard routine in pre-employment or periodic medical examinations. The classical tests - possibly with some changes - can prove very useful, as they afford a simple means of assessing the physical capabilities of adolescents and following up their subsequent development. Where particular problems arise, however, because certain deficiencies have been found or on account of the nature or conditions of the work to be performed, the full battery of cardiorespiratory function tests can be carried out in order to gain an over-all picture of the adolescent's aptitude for effort.

Respiratory Apparatus and Lung Function

During adolescence, the thoracic cage grows considerably, especially in boys. The respiratory parameters undergo a change. The vital capacity, which has been increasing since childhood, undergoes a fairly substantial increase in girls between 12 and 13 years of age and boys between 13 and 14 years of age, after which the increase continues, but at a slower rate. From the age of 14 onwards, considerable variations between individuals are to be observed, higher values of vital capacity being reached in boys than in girls. Maximum ventilation per minute increases until the age of 14 years, after which it levels off, and is greater in boys than in girls. A similar difference also exists in respect of maximum expiratory volume per second. Respiration, which was primarily abdominal in the child, assumes a thoracico-abdominal character, principally in boys.

In evaluating the adaptation of lung ventilation to effort, three main aspects must be considered, namely, what are the available pulmonary volumes, how are they used, and what is the extent of gas exchange during effort? Vital capacity is easily measured, and it is interesting to follow its evolution over a period of time, but one must not forget that this is a purely biometric parameter, from which no deductions can be made regarding aptitude for effort. The maximum volume expired per second and maximum ventilation per minute may also be measured. Special techniques are required, however, in order to measure the other pulmonary volumes.

While sensitivity to infections of the upper respiratory tract is high during puberty and the period preceding it, it appears to decrease in adolescence, and even to be replaced by a measure of resistance to such infections.

The habit of cigarette smoking is so widespread among young persons that it is not unusual to find adolescents suffering from smoker's cough, or even from bronchitis which may become chronic in time. It is essential that adolescents be given appropriate health advice and realistically informed about the potential hazards of this habit.

The highest importance attaches to the systematic detection of tuberculosis. This disease, which is still encountered in all countries, albeit to a greater or lesser extent, remains a menace, even in areas where it has become much less prevalent. In some countries, tuberculosis is in the forefront of all occupational health problems and carries with it a high rate of morbidity and mortality.

The detection of tuberculosis is especially important during the prepuberal period and in adolescence, for these stages of life appear to be characterised by a particular sensitivity to this disease. The reason may lie in the intense changes to which the body is subject at this period, and in the substantial nutritional demands which accompany them and which are not always satisfied, with the result that resistance is lowered. Furthermore, initial contact with the tuberculosis bacillus and the change from a negative to a positive reaction to tuberculin (Mantoux conversion) now occur later in populations who are effectively protected against tuberculosis, as is shown by the trends in the tuberculination indices; thus, the first contact will frequently take place during adolescence, that is to say, precisely at a time when resistance is lowered.

The problem of tuberculosis prevention is above all a public health problem, and preventive measures taken in respect of young workers in undertakings must fit within a co-ordinated over-all programme. Emphasis must be laid on the importance of periodic checks of tuberculin sensitivity, and of X-ray examination of the thorax. However, it must not be forgotten that tuberculosis is not restricted to the lungs, although they are the most frequent site. A lung X-ray examination will detect tuberculosis of the lungs, but not elsewhere; if it is negative, this does not exclude the possibility of a developing tuberculosis of the kidneys, gonads, abdomen, bones or joints. If the campaign is aimed at all kinds of tuberculosis, the subject's medical history must be carefully studied and he must be given a complete clinical examination.

Pneumoconiosis is a consequence of the prolonged inhalation of harmful dusts. The longer the exposure, the greater is the risk of serious disease; it follows that exposure should be delayed as long as possible. Young persons tend to display a high level of physical activity (often somewhat disorganised), which increases the lung ventilation and, with it, the likelihood of inhaling dust. Indeed, cases of very rapidly developing pneumoconiosis - silicosis in particular - sometimes occur in young workers; these cases may even be so acute as to be tantamount to poisoning. Young workers who may be exposed to a pneumoconiosis hazard must therefore be given an especially careful check-up and periodic examination.

Asthma is a very delicate problem, as this disease may jeopardise the occupational future of a young worker. The physician may be faced with an asthmatic adolescent, or with one in whom - as often occurs - an early asthma disappeared at the end of childhood or the advent of puberty, and the question arises in what measure this history may be responsible for a heightened sensitivity to work-related respiratory diseases (e.g. occupational asthma, pneumoconiosis), and to what extent certain occupations might be contra-indicated as a result. Handling these problems successfully requires experience, common sense and an intimate knowledge of conditions at workplaces.

If the asthma is congenital, the young person should be preserved from contact with new allergens, but in practice this is not always possible. He should be advised to avoid occupations entailing a risk of contact with allergens known often to affect the respiratory tract, which may be dusts of vegetable or animal origin or any one of a variety of chemicals. He should also avoid work involving possible exposure to irritating gases or vapours, or frequent exposure to chilly conditions. The choice of an occupation should be made with these restrictions in mind, and appropriate vocational guidance given as necessary for initial placement or redeployment.

If the young worker had no previous history of asthma, but an asthmatic condition was complained of or noticed on the occasion of a periodic examination, the question then arises, is it really asthma? and if so, is it work-related? Questioning will bring out the most important information; patch tests, and especially aerosol inhalation tests, may be useful. Very frequently it will be necessary to withdraw the adolescent from work in which the suspected allergen may be encountered, especially since if exposure is allowed to continue it may lead to polysensitisaticn. In his employment, occupations in which other very active allergens may be encountered should therefore be avoided, and appropriate vocational guidance, backed up if necessary by training or retraining, should be given.

Blood-forming and Glandular Systems

As an adolescent's body develops, the blood circulation increases in volume and must adapt itself to the new conditions arising. The total haemoglobin content increases regularly until the age of 20 (girls) or 22 (young men), after which it levels off. A significant rise in haemoglobin level occurs between the ages of 11 and 15. During the prepuberal period, this level is identical in boys and girls, but after puberty, it is higher (by about 1.5-2g per cent) in the former.

Young workers are known to be particularly prone to occupational accidents, as well as to serious road accidents (including commuting accidents) occurring towards the end of adolescence, when they have just learned to drive a car, or acquired a light or heavy motor cycle. It will accordingly be advisable for every young worker to have his blood group taken and to carry a card with this information on him always.

Anaemia may adversely affect adaptation to effort and to work, owing to the reduced capacity of the blood to transport oxygen. It may be responsible for dyspnoea on effort, palpitations or undue fatigability.

It is not exceptional to find anaemia in young persons, for example at a pre-employment examination. Such anaemia is frequently hypochromic and hypoferric in nature and may be due principally to inadequate iron intake or nutrition, or to absorption difficulties. Loss of blood may also be responsible (e.g. hypermenorrhea or metrorrhagia in young girls), as may iron capture by the reticulo-histiocytary system in chronic infections, or deficient bone marrow iron metabolism (thyroid disorders, vitamin C deficiency, lead poisoning). A doctor who finds anaemia during a periodic examination should make sure that it is not the result of exposure to a poison at the workplace.

The blood leucocyte count may be abnormally high or low, in which case additional examinations will be necessary. If this condition is found at a periodic examination, the possibility of poisoning should be investigated. Disorders of blood coagulation and haemorrhagic syndromes may be found, and the reason will have to be established, a possible toxic origin not being disregarded. It must not be forgotten that myeloid leucosis mainly afflicts young persons.

The blood-forming tissues are particularly active in adolescence, especially during the prepuberal and puberal periods; it is accordingly logical to consider that they will then be more sensitive to the effect of certain poisons, although there may not always be proof of this. One must be mindful of the special threat to the blood-forming system which might result from the effects of ionising radiation, if exposure thereto before the age of 18 was not prohibited, of the particular sensitivity to benzene which appears to exist, especially in early adolescence, and of the apparent special sensitivity of adolescents to lead. Exposure of young workers to these hazards is prohibited by a large number of laws and regulations.

In young persons, the examination of the glandular system should not be omitted, since various affections involving polyadenopathies or localised adenopathies occur fairly frequently around the time of adolescence (e.g. infectious mononucleosis, sarcoidosis (Besnier-Boeck-Schauwman disease), Hodgkin's disease, etc.). The presence of a localised adenopathy, particularly in the region of the neck, should never be taken lightly, as it may be symptomatic of conditions which require to be tackled quickly and in which a biopsy must be taken without hesitation. The occupational health specialist may be faced with this problem, either because the young person concerned attaches no importance to it, or because he is afraid of going to a physician for treatment.

Aptitude for Effort, Working Capacity and Energy Expenditure

The ability of a given person to exert a given effort or perform a given physical task depends on a number of factors, resulting from the integration of a whole series of systems and functions which are subject to variation, in particular with the age of the subject. Of these factors, the following in particular must be mentioned:

- The conditions of the muscles and of the neuromuscular functions in general must be considered (muscular development, strength and endurance, motor co-ordination; training and techniques acquired).

- The available energy corresponds to the amount of energy that the anaerobic or the aerobic metabolism can supply. A person's capacity to provide energy through the aerobic metabolism depends in particular on his capacity for transferring oxygen in the air to the muscles, and is affected by the following factors: lung ventilation, alveolar and capillary diffusion, haemoglobin level, volume of blood, cardiac output (systolic output and heart rate), blood circulation rate, tissue perfusion characteristics (pressure, adaptation of the circulation, with preferential perfusion of active muscular zones, adaptation of local capillary circulation), diffusion of oxygen from the capillaries to the muscle cells, and venous return. The aerobic capacity also depends of the way in which the oxygen supplies will be used within the cells, that is to say, on the efficiency of the enzyme systems controlling production and use of adenosine triphosphoric acid.

- The importance of psychological aspects (interest, motivation, will) must certainly not be underestimated. Appropriate techniques can be used to make work more effective while eliminating unnecessary energy expenditure, leading to maximum efficiency achieved with minimum effort. During adolescence, a period of growth and maturation, the factors just mentioned vary continually, but not necessarily simultaneously, with age. The condition of the muscular system and respiratory, cardiac and circulatory adaptation are also subject to variation. At this stage of life, psychological aspects, as has already been seen, take on particular significance.

When mention is made of aptitude for effort or for work, the first reaction is that this is a simple concept, which corresponds to a person's ability to make efforts or perform physical work. The assessment of this ability does not appear to pose any problems, since in daily life one has no difficulty in evaluating subjectively the performance of persons whom one encounters or knows: one notices that one can climb three flights of stairs without being out of breath, whereas one's neighbour must stop for a breathing spell; during sports activities, it is seen that some people perspire freely, while others do not; a neighbour can work at his garden all day without being tired, while it takes others two days to get over such exertions. Through such repeated observations, one learns to recognise that such and such a person is capable of performing heavier physical work than another person, or can perform the same work with less fatigue, and that this work which must be done may be entrusted to that person without giving rise to particular difficulties, notable fatigue or physical overstress. However, when one seeks to be more specific and to find out what this subjective evaluation - this impression, as it were, of physical fitness - is really based on, matters become more complicated.

In practice, this evaluation is based on a number of criteria, such as the appearance of the person during effort, his place of work, regularity and ease or otherwise of movement or respiration, extent of perspiration, and signs of fatigue; but if one seeks an objective evaluation, this must be founded on measurable criteria, and it is no easy matter to select these and give them relative ratings or "weights" with respect to the others. The question then arises as to which criterion will in fact be most representative of this "aptitude" which is to be evaluated. Even assuming that the ideal criterion is found, there is no hope of measuring all its variations for all kinds of effort of varying intensity and duration, and a specific kind of effort must be selected. A further question then arises, namely, to what extent the evaluation of aptitude arrived at for one kind of effort will hold good in other circumstances, and whether what has been measured is merely aptitude for a specific effort, or does correspond in fact to the subject's general capacity to undertake physical efforts in good conditions from his own point of view and that of the work to be done.

The choice of a criterion of aptitude for physical effort is a difficult matter. Measured cardiac and blood circulation changes might be used, but in doing so there should be no confusion between characteristics of cardiovascular adaptation to effort and aptitude for effort. While this measurement is no doubt essential, it does not represent the whole criterion, as has been seen; moreover, the result might be to over-emphasise the importance of training, compared to physical capacity _per se_. Fatigue might also be used as a criterion of physical aptitude, the person most fit to perform a given task being deemed to be the person who is least tired by it, or who does most work or displays the greatest activity for a given degree of fatigue. But here, one immediately stumbles against the unsolved problem of the evaluation of fatigue; sometimes, moreover, such a course might result in too much importance being attributed to psychological and highly subjective factors, or to more importance being attached to a person's ability to work without becoming tired by using certain efficient working methods than to the evaluation of aptitude for physical effort.

The kind of effort to be made is also important, since there is a great difference between the body mechanisms involved and adaptations required in making an intense effort for a few minutes or hours, and in performing physical work such as is normally encountered in occupational life. Thus, the measurement of maximum oxygen consumption per minute, which is a very good yardstick of physical aptitude for efforts of short duration, might not be a very accurate test of aptitude for occupational work, the intensity of which varies without reaching any great heights, but which must be continued throughout a working day of eight or nine hours and resumed on the morrow, the day's fatigue being completely eliminated by nightly rest.

These considerations go to show that it is extremely difficult to arrive at an objective definition of the concept of "physical aptitude" or fitness, which should enable one to predict confidently that a certain person will be able to perform a certain kind of work, which he has never done before, without any form of overstress. They show that in practice, evaluations of physical aptitude can only hope to get somewhere near this objective. Clearly, moreover, if it is difficult to measure an aptitude, it will be even more difficult to define an incapacity. If one adds to that the fact that a ruling of unfitness may have very grave consequences on the human plane, it will be possible tc grasp the full complexity of the problem which the occupational health specialist may have to face in attempting to answer an apparently simple question.

Many tests which seem likely to provide a good evaluation of fitness for effort or work have been studied. Certain of these are worth listing as being particularly suitable; they are:

- measurement of maximum oxygen consumption per minute (Astrand);
- measurement of work in kilogramme-metres per minute, performed at a heart rate of 170 beats per minute ("working capacity": Wahlund, Bengtsson);
- measurement of oxygen consumption per minute for work performed at a given heart rate;
- study of increase in heart rate as a function of increased effort and of the coefficient (Leistung-pulsindex) by which these two variables are related (Rutenfranz, Hettinger).

Most of these methods, however, require equipment which is not usually available to industrial medical services, as well as an expenditure of time which industrial physicians are not always able to afford. Although it is obviously desirable that the technique used should provide information as closely related as possible to general aptitude for physical effort, this does not mean that less accurate methods are entirely valueless. Even if the technique used falls a good way short of scientific perfection, it must not be forgotten that by dint of personal experience, a doctor will learn to interpret it correctly and thus gain a good idea of the physical capacity of the persons thus examined. For example, a physician who is familiar with the effort test imposed and knows that such and such persons who obtain such and such results adapted well to certain kinds of work, while other persons with other results did not, who is familiar with the workstations concerned and has a good idea of the "physiological load" that they represent, will in time be able to estimate very accurately whether a given person can be employed at a given workstation without harm to his health and well-being.

It is interesting tc compare the evolution of physical capacity for effort in adolescents and in adults. As maximum oxygen consumption per minute is agreed to be one of the best yardsticks of a person's capacity for intense physical effort, since it embraces all the factors affecting the transfer and use of oxygen, most research has been made on this basis. Using this criterion, it is possible to determine the maximum energy that a person can develop per minute; this is because, whereas for efforts of less than two minutes' duration, the necessary energy is essentially obtained anaerobically, for more protracted efforts, the aerobic metabolism becomes the essential source of energy, and there is a close correlation between oxygen consumed and energy released.

Before puberty, there is no significant difference in maximum oxygen consumption per minute as between boys and girls. After puberty, it is lower in girls, the figure being only 70 to 75 per cent of that for boys. In both sexes, the highest values are found towards the age of 18 to 20. There are constitutional variations between individuals, and other variations which may be related to training, which can raise the maximum oxygen consumption per minute by some 10-20 per cent; the effect of training is greater during adolescence, between the age of 10 and 20, than in adulthood. These findings might suggest that maximum aptitude for intense effort is reached towards the age of 18 to 20, but it is observed on the contrary that during intense efforts requiring a certain endurance, the best performances are achieved by persons 25 to 30 years old. This may be due to the fact that with training, the maximum oxygen consumption values can be maintained between the ages of 20 and 30, during which time ways of using the energy produced more judiciously can moreover be learned.

While it is interesting to be acquainted with the evolution of capacity for intense physical effort and maximum energy expenditure, it is no less interesting to be able to evaluate the workloads that should not be exceeded in the case of adolescents.

Studies of the nutrition of workers engaged in heavy work in industry, mining or agriculture, made by the Occupational Physiology Institute in Dortmund (Federal Republic of Germany) have shown that these workers rarely consume more than 4,300-4,400 kcal per day (average calculated over a number of years). Now, it is considered that to maintain his weight at the same level, a normal, healthy man requires a daily intake of 2,400 kcal, as long as he is performing no substantial physical work (these 2,400 kcal are used as follows: 1,600 kcal approximately for the base metabolism, 200 kcal for the specific dynamic action of foodstuffs, 200 kcal for losses due to non-use of foodstuffs, 200 kcal for maintaining a very reduced level of activity, e.g. getting up, washing, dressing and moving around indoors). It thus appears that the performance of heavy work entails an energy expenditure of some 2,000 kcal daily in a working day of eight hours; so high a level of energy expenditure is reached only by 15-20 per cent of workers in heavy industry, and is rarely exceeded. It is consequently considered that 2,000 kcal daily is a limit which must not be exceeded; this figure is equivalent to an energy expenditure of 4.2 kcal/min. (The limit is set at 4 or 5 kcal, according to the authors.)

Similar studies have been made of the ability of young persons to perform work involving substantial physical effort. It has been estimated (Hettinger) that if the maximum daily expenditure of calories is 2,000 kcal in men and 1,400 kcal in women, it must not exceed:

- in boys: 1,400 kcal at age 13-14 years (70%);
  1,600 kcal at age 15-16 years (81%);
  1,750 kcal at age 17-18 years (87%);
  1,900 kcal at age 19-20 years (95%).

- in girls: 1,200 kcal at age 13-14 years (60%);
  1,300 kcal at age 15-16 years (65%);
  1,350 kcal at age 17-18 years (67%);
  1,400 kcal at age 19-20 years (70%).

If one considers that the daily energy expenditure corresponds to up to 1,000 kcal for light work, 1,000-1,600 kcal for medium work, 1,600-2,200 kcal for heavy work, and over 2,200 kcal for very heavy work, it follows that until they reach the age of 16, young workers should not be employed on heavy work. The energy expenditure of 1,600 kcal required by a 16 year old adolescent to perform medium work corresponds to an expenditure of 2,000 kcal in an adult. While the maximum value for energy expenditure per minute during a working stint of eight hours is about 4.2 kcal/min for adults, it is evaluated at about 3.3 kcal/min for adolescents (Hettinger). Since industrial work is planned for adults, it is possible that this limit may be exceeded, and the especial importance of providing for additional and longer work breaks for young workers, according to need, will therefore be understood.

On the occasion of the pre-employment examination and the placement of young workers, as well as of periodic examinations, the occupational health specialist will have to study these questions relating to the aptitude of young workers for effort, and to ascertain that the work entailed does not exceed the physiologically recommendable norms. It must also be remembered that, while work can be arduous on account of the energy expenditure that it entails, other factors must also be considered, such as adverse environmental conditions (high ambient temperature and humidity, radiant heat, noise, vibration, dust, poor lighting, etc.) and the existence of static components in the work done (e.g. holding a heavy tool, carrying heavy loads, maintaining a fixed working posture, etc.). An idea of the arduousness of the work can be obtained by measuring the increase in heart rate and comparing it with the increase which would be observed during dynamic work performed in good environmental conditions, for the same energy expenditure. It is generally agreed that the increase in the heart rate should not exceed the normal rate at rest by more than 40 beats a minute. While this value is regarded as being the same in adults as in adolescents, it will be noted that in young persons it is reached more quickly and for lighter work, just as a greater increase in heart rate is required for the same increase in blood circulation, as already pointed out.

Problems of Eyesight

A test of eyesight, with a diagnosis of any deficiencies found, is an essential part of any pre-employment medical examination, and the importance of subsequent regular check-ups must be stressed. The number of occupations in which good eyesight is mandatory is on the increase (substitution of dial reading at a control station for manual labour, and of mechanical handling for manual handling, increasing emphasis on supervision and quality control, etc.). Moreover, good eyesight is important from the safety standpoint.

The eyesight test often consists simply in checking vision with test types at a distance of 5 metres. This is only a first step, and the physician should not rest content with it, especially since there are simple and rapid methods available for checking visual function satisfactorily. When the person is to be employed at a workstation where good eyesight is specially important, additional examinations should be made. When an anomaly or deficiency is noted, it should be diagnosed and appropriate action taken by an opthalmologist.

The subject should be examined for refractory disorders (hypermetropia, myopia, astigmatism). A simple method consists in checking the sharpness with which each eye perceives distant objects; if it is insufficient, the young person may be short-sighted or suffering from another defect of vision. If perception is normal, this means that the subject has normal vision or is long-sighted, the diagnosis being established by the use of convergent lenses of 1.5 dioptres; if distant objects are still perceived sharply with these lenses, then hypermetropia is present, while a loss of sharpness indicates normal vision. Astigmatism, depending on its extent, may or may not entail a loss of visual acuity; it should be checked by the appropriate tests. Deficiencies of eyesight due to refractory disorders can be compensated by corrective lenses. Visual acuity should be tested again after suitable correction has been applied.

In the placement of short-sighted persons, account should be taken of the fact that they perceive objects close at hand best. However, some correction may be necessary even for near vision in order to permit a normal use of convergence. As short-sighted persons are more prone than others to detachment of the retina, particularly where there are evolutive lesions of the fundus oculi, they should not be employed at workplaces where the body, and more especially the head, is exposed to vibration or jolting (pneumatic hammers, tractor driving, etc.).

In placing long-sighted persons, account should be taken of their defective near vision and of the difficulties that they may experience in consistent close work (e.g. sewing and fine-drawing in the textile industry). Consideration should also be given to the possible sensitivity of astigmatic persons to dazzle.

When corrective spectacles have to be worn, account must also be taken of the additional accident risks entailed and of the possibility of the lenses becoming steamed up in work in steamy conditions (e.g. wool washeries) or in passing from a cold atmosphere to a warmer and humid one, not forgetting, however, that in some jobs, spectacle wearers have the advantage of some measure of permanent protection (e.g. from foreign objects in the eye). Contact lenses offer both advantages (reduction of accident hazards, in some cases, improvement of myopia) and disadvantages (they are not always well tolerated). Reference should be made here to the increasing use of corrective lenses of hard plastic which are characterised by good mechanical resistance and transparency. Some undertakings encourage their workers to acquire such lenses.

Reduced perception of distant or nearby objects may be due to a number of causes. These include, in addition to the refractory disorders already referred to, a decreased transparency of the ocular media (affection of the cornea, thickening of the crystalline lens, or complications following exposure to infra-red or ionising radiation), as well as lesions of the retina or of the nerves. Possible other causes also include poisoning by certain substances or smoking to excess; the loss of visual acuity may also result from a squint accompanying untreated far-sightedness, or due to some other cause. Whenever defective vision is noted at a periodic examination, the industrial physician should always consider the possibility of poisoning, e.g. optic neuritis induced by methyl alcohol, trichloroethylene, etc.; conditions at the workplace should be checked and may yield very useful information to the specialist. Subjects displaying a sudden or marked decrease in visual acuity should be referred immediately to an ophthalmologist. When vision is nil or less than one-tenth for the better eye of the two, the subject is regarded as blind. When it amounts to between one-tenth and four-tenths for the better eye after correction, the subject is suffering from partial loss of sight (amblyopia).

Determination of the visual field and any anomalies is important, particularly if the young person is to be employed at certain workplaces requiring good peripheral vision, as in the operation of vehicles or cranes, the handling of voluminous objects, or work with dangerous machines or in dangerous or crowded areas. For accuracy, of course, the field of vision should be checked with appropriate instruments (e.g. Goldmann-type perimeter) and this may not always be included in the routine examinations performed by an occupational health service. Nevertheless, it is always possible to check the extent and symmetry of the field of vision very simply by asking the subject to look at a given point and at the same time to indicate when he can see a finger which the examiner, positioned behind the subject, interposes in his visual field from different directions. The individual field of vision of either eye can also be checked separately in an equally simple manner.

A reduced field of vision may result from a variety of causes, and should always be accurately diagnosed. If it is first discovered at the pre-employment examination, precautions should be taken in placing the subject, for his own safety's sake and for that of others. If it is first found during a periodic examination, the possibility of an occupational origin should not be overlooked, such as incipient poisoning (by carbon monoxide in particular).

Binocular vision is frequently more important than one would imagine to be the case in occupational life, and particular attention should be devoted to it at the medical examination, principally when certain specific jobs, such as the operation of vehicles and various machines, are involved. An adequate check can be made simply and quickly, for example by means of a stereoscope or diploscope, or by using Worth's test, etc.

Good colour vision is essential for the correct perception and comprehension of numerous safety codes and signals used in aviation, rail and road traffic, industrial safety practice, etc., and is necessary in a large number of occupations (e.g. work with dyestuffs, use of colorimetric reactions in chemistry, painting and decorating, etc.). Tests for colour blindness should therefore also be given by the occupational health specialist, the more so since this can be done very simply and rapidly, for example using Ishihara's plates. Although the aim is essentially to detect congenital colour blindness, if an acquired anomaly should be found at a periodic examination, the possibility of a toxic origin should be investigated. Deficiencies of colour vision must be taken into account in considering the placement of young workers in certain jobs, as well as in vocational counselling.

Deficiencies of visual adaptation to variations in ambient light may also be present. For example, the subject may be unable to see and lose his sense of direction in passing from a lighted area to one in darkness, or may be very sensitive to dazzle. The importance of such a deficiency from the safety standpoint and for employment in certain jobs will be easily understood. Special equipment is necessary in order to check up on such deficiencies; however, useful information can be obtained by questioning. They may be congenital and hereditary, or acquired, in which case they may stem from some general affection or affection of the eyes, or may possibly have a toxic origin, which should be investigated.

The occupational health specialist may also be concerned with disorders of ocular mobility, convergence or accommodation, as well as with the problem of squinting and its effects on the eyesight.

Medical examinations may show up a whole series of affections of the eyelids, lacrimal glands, conjunctiva, cornea, iris, pupil, etc., which should be treated by a specialist. They need not be gone into here. Reference should however be made to the xeroma which exists in certain countries and results from a lack of vitamin A in childhood. This condition first develops in infants two to three years old, and may result in total blindness; the works doctor may find sequels of this serious affection in some adolescents. Young persons also appear to be rather prone to allergic conjunctivitis, and in such cases, a possible occupational origin should be sought.

In many countries trachoma remains an extremely serious problem, despite all the efforts made to combat it and the extensive use of effective ophthalmological medication. Contamination is particularly frequent in childhood, so that the problem exists from infancy, but very frequently, in endemic areas, a developing trachoma, often previously unrecognised as such, may be found in young workers. In such cases treatment must be given and medical supervision, over a protracted period, arranged. Industrial physicians can make a very useful contribution at the level of the undertaking to public health programmes aimed at diagnosing and treating trachoma. The threat posed to eyesight by this disease, which may lead to total blindness, is a source of serious problems of job placement, vocational guidance and retraining of workers affected.

At medical examinations, the occupational health specialist must not fail to draw the attention of young workers to the precautions that they must absolutely take to preserve their eyesight, and in particular to the importance of using safety goggles for protection against dust, flying particles, acids, and infra-red and ultra-violet radiation, etc. He must stress the fact that any eye injuries must be treated quickly and effectively, since otherwise even apparently insignificant causes can give rise to very serious and possibly incurable lesions. Instilling a fear of blindness is, in fact, a salutary method of overcoming the dangerously careless approach that is all too often adopted towards eye protection.

Hearing Problems

Hearing problems in adolescence are perhaps more frequent than is generally supposed; often, they are overlooked simply because hearing has not been adequately checked. Whereas the person concerned will notice a loss of hearing in both ears fairly quickly, he is less likely to notice a moderate loss affecting one ear only, and many such cases will be discovered only by medical examination. According to certain statistics, one young person in twenty-five suffers from hearing deficiencies. Yet a good sense of hearing is necessary in many jobs. For work to go forward normally, orders must be understood, and the normal or abnormal noise produced by machinery represents essential information. Warning signals and unusual noises heralding a dangerous situation must be perceived, even in a noisy environment. Thus good hearing may be essential both for the transmission of information and as a safety factor.

Frequently, all too little consideration is paid to hearing tests in occupational medical examinations. Although it is hardly possible to make all young workers undergo a complete audiometric examination, which is a long procedure requiring special instrumentation and a soundproof room, this is no excuse for doing nothing, and the medical check-up should normally include an otoscopic examination and an assessment of hearing by conventional acoustic methods (loud speech, whispering, ticking watch, etc.). The possibility of using certain simplified audiometric methods, which can be performed quickly without costly instruments, should also be considered.

A decrease in monaural or binaural hearing sensitivity may be due to a variety of causes, some of which will be immediately obvious. Frequently the ears will be found to be blocked by wax, especially where hygiene leaves something to be desired. The eardrums may have shrunk as a result of otitis, they may be perforated, or a chronic discharge may be present; there may be a history of meningitis or head injury. The occurrence of hearing loss in persons exposed to noise may point to incipient occupational deafness. Fairly often, however, the cause will not be immediately apparent, and additional examinations will have to be made by a specialist in order to arrive at a precise diagnosis and determine the correct treatment.

If a hearing deficiency is found at the pre-employment medical examination, the dual problem then arises of determining on the one hand whether the deficiency is compatible with normal work in conditions of sufficient safety for the young person and his workmates, and, on the other hand, whether environmental conditions at the workplace might not tend to aggravate the existing loss. If orders or safety signals are mainly aural, as, for example, in winding operations in certain mines, a person with a hearing defect may be unfit for employment. Also, work in certain noisy surroundings may be contra-indicated for persons with a hearing loss, as they are often more sensitive than others to harmful noise. The idea that workers with hearing deficiencies can be employed at noisy workplaces because they will be less affected than their workmates is entirely mistaken and should be eradicated. It holds good only in the case of persons who are totally deaf and can therefore suffer no further hearing loss. Workers using hearing aids should be employed at quiet workplaces, as otherwise the noise amplified by the hearing aid could quickly destroy whatever sense of hearing remains.

In some countries, workers scheduled for employment in certain particularly noisy locations are required by law to undergo pre-employment and periodic medical examinations. These must include a complete audiometric examination; hearing at frequencies of 4,000 Hz and over must be explored with particular care, in order to pinpoint any deficiency occurring at around that frequency, which will suggest a diagnosis of occupational hearing loss. A study of the medical history will provide information on the possible general effects of noisy environments.

Industrial medical services would be well advised to keep a special check on the hearing of young workers, and to subject as many young persons as possible to such supervision, in view of the likelihood that adolescents are specially sensitive to noise, not only as regards its effects on hearing, particularly at the higher frequencies, but also as regards its general effects.

When signs of incipient occupational hearing loss are noted, the industrial physician will decide on the best course to follow (withdrawal of worker from noisy environment, more effective use of personal protective equipment under medical supervision, etc.). At medical examinations, he must not fail to draw the attention of young workers to the importance of using personal protective equipment for hearing protection consistently and correctly.

Problems Affecting the Central Nervous System

Motor and Perceptive Faculties

In adolescence, while the basic capability of achieving movement has long since been acquired, a whole series of motor aptitudes continue to be developed, such as speed and accuracy of movement, muscular co-ordination, motor automatism and manual dexterity. The perception of space and time is reinforced. This period of life seems to be particularly favourable for acquiring good motor co-ordination, automatism and techniques. Apprenticeship seems to go through more quickly and to produce more lasting results than in adults. This is the best time for training in occupational and sporting techniques.

The processes by which the mental faculties develop and mature influence the manner in which information reaching the central nervous system is perceived, analysed and integrated. They also affect the individual's response to these external stimulations, and his motor or psychomotor reactions to them. As a consequence of the development of intelligence, affectivity and imagination, proprioceptive perceptions and autonomic function sensitivity may reach hitherto unknown levels of intensity during adolescence.

Autonomic Nervous System

In adolescents, the autonomic nervous system is particularly labile, and this condition might be said to be characteristic of this stage of life. Some authorities relate it to the fact that in childhood the parasympathetic generally predominates, whereas in adolescence, as in adult life, the sympathetic is frequently somewhat predominant.

This lability of the autonomic nervous system is apparent particularly in the intensity of reaction to emotions, including peripheral vasomotor reactions (blushing), palpitations, marked increase or drop in blood pressure, perspiration, etc.

Mention has already been made of the frequent occurrence of autonomic disorders affecting the cardiovascular system; very often, also, these disorders are responsible for digestive symptoms, which may be very unpleasant and disturbing, but are as a rule not really serious, although if they become chronic they may develop into psychosomatic illnesses finally giving rise to organic lesions (gastric ulcer, haemorrhagic rectocolitis, ulceromembranous colitis, etc.).

When disorders of the autonomic nervous system are found, the question should be asked whether the mental factors responsible for them do not originate at the workplace. Are these disorders due to excessive fatigue, to poor adaptation of the subject to his work (lack of interest, monotony, too rapid a pace of work, etc.) or of the work to the worker? Is he perhaps afraid of the work, but unwilling to say so (dangerous machinery, underground work, claustrophobia, etc.)? Is there a lack of integration in the undertaking, an underlying climate of conflict with workmates, or a feeling of inferiority? By taking an interest in these problems and seeking to determine the true cause of the disorders noted, the occupational health specialist can often be of considerable service to young workers and may be in a position to supply valuable data to the physician who is treating their condition.

Left-Handed Persons

Left-handed adolescents are normal persons characterised by a tendency to use the left-hand side of the body (hand, upper and lower limb, eye) in preference to the right as is habitual in most persons. In true left-handed persons, the right cerebral hemisphere is dominant, whereas in the majority of persons the contrary holds good. A distinction must be made between true left-handedness and behaviour acquired for example as a result of education by a left-handed person or of "mirroring" the behaviour of adults in early childhood. Left-handedness may be total or partial.

In the adolescent, ordinary behavioural training has already taken place. The problems that arise are no longer the same as in childhood and are perhaps no longer so important; thus action to counter left-handedness in adolescents will not have the same consequences as in young children. Nevertheless, it is preferable not to attempt to re-educate a markedly left-handed person, especially since in certain jobs performed by two or more workers (as in carpentry, certain assembly operations or surgery, for example), there may be a distinct advantage in being left-handed. It also appears that during vocational training, little difficulty is experienced in teaching left-handed young persons to be ambidextrous, which is a very useful attribute.

No serious problems appear to arise in undertakings in which all the installations are designed for right-handed persons, no doubt because left-handed workers adjust to this situation or adapt their tools or machines to suit their convenience; it is noteworthy that left-handedness is rarely invoked as a cause of accidents.

The occupational health specialist may be concerned with the question of left-handedness in certain circumstances, for example when a request is made to redesign a workplace for a left-handed worker, in which case, he will first have to verify whether the worker is really left-handed and to what extent, and then study the measures to be taken. Also, when faced with an unusually clumsy young worker, who has difficulty in learning certain occupational tasks and movements or in distinguishing the left side from the right, he should investigate whether left-handedness is not at the root of these problems.

Accidental Injuries and Other Affections

Young persons are prone to accidents, and perhaps more especially to those involving head injuries, as is apparent from road accident statistics, especially those involving motor cycles. Motor cyclists should be obliged to wear crash helmets, and industry could usefully participate in campaigns towards this end, in view of the increasing frequency of commuting accidents. Reticence, if any, on the part of young persons to wear crash helmets is a matter of sheer habit or of fashion and can be overcome by judicious propaganda.

Recovery from cerebral concussion is generally good, and subsequent protracted headache or other syndromes developing into a chronic condition appear to occur less frequently than in adults. In view of the significant part played by psychic factors in the aggravation of post-concussive conditions, particular care must be taken to avoid anything which might encourage young accident victims to lodge exaggerated claims for compensation (which are moreover usually inspired by their parents), as there is always a risk that some form of sinistrosis may ensue.

Adolescents, it has been said, are particularly sensitive to tuberculosis, and a similar sensitivity exists to tuberculous meningitis; periods of marked fatigue or intensified activity may result in lowered resistance to this disease. In particular when the works doctor is dealing with young workers living in collective accommodation made available by the undertaking, an outbreak of meningococcal meningitis (epidemic meningitis) will entail problems of isolation and prevention which must be solved in all haste, in view of the speed with which this disease can spread in groups of young people.

A large number of products used at work may affect the working of the central nervous system. This is true in particular of solvents (benzene, toluene, xylene, trichloroethylene, perchlorethylene, etc.) and of all fat-soluble substances in general. In undertaking periodic examinations of workers exposed to these products, one should be on the lookout for any slight signs of cerebral dysfunction or neurological symptoms. Impressions of inebriation, lack of balance, unsteady gait and sleep disorders are suggestive of acute poisoning, the causes of which should be sought at the workplace. Workers exposed to mercury or manganese should be systematically checked for any signs of tremor. Headaches, particularly those occurring systematically after a time spent at the workplace, are suggestive of carbon monoxide poisoning.

Mention must also be made of certain diseases which the occupational health specialist may be the first to detect, because they are initially insidious and make themselves felt only in a certain lack of dexterity or momentary loss of balance which may be a cause of anxiety or lead to falls or occupational accidents. Examples are Sydenham's chorea, which attacks adolescents in particular, and multiple sclerosis, which appears at the end of adolescence and in young adults. Many other affections can be cited in which the medical history shows that it was at the workplace that the patient first became aware that something was wrong (brain tumours, aneurisms, etc.); it can therefore be the case that the works doctor is the first to be consulted about the small initial symptoms of such affections.

Psychological Disorders related to the Maturing Process

Attention must be drawn immediately to the importance of distinguishing between psychological disorders arising out of the maturing process, which are frequently referred to by the term "juvenile crisis" or "crisis of adolescence", and disorders corresponding to neuroses or psychoses. This is no easy matter, however, since the initial behavioural symptoms are identical or very similar, and frequently the final diagnosis will depend on the course taken by the disorder.

Problems of maturation, neuroses and psychoses are all common occurrences in adolescents, owing to the personality changes which are taking place and the attendant instability. At this time, pathological psychic structures (often pre-existing) become patent.

Although one may come across obsessional neuroses, phobias and hysterias, etc., the essential problem, which will arise fairly frequently, is that of the early diagnosis of schizophrenia and of its minor manifestations. When the presence of a manic-depressive syndrome, schizophrenia, or phobic or obsessional anxiety neurosis is suspected, the subject should be directed to a psychiatrist for diagnosis and treatment in depth. Some forms of anxiety neurosis may be related to difficulties encountered at the start of occupational life, while certain phobias may be related to work or the circumstances in which it is performed (fear of machines, bosses or occupational accidents or diseases). The diagnosis of hysteria should never be superficial, but should be made only after everything has been done to discover a possible somatic cause. Descriptions have been given of hysterical behaviour at workplaces, principally affecting groups of women workers, especially young workers; in such cases, a check should first be made whether conditions at the workplace may be responsible for the behaviour observed. Thus, symptoms of malaise, accompanied by falls or fainting fits, may be due to heat, poor ventilation, escaping gas or poisonous substances (carbon monoxide, solvents, etc.). The occurrence of collective hysteria is explained by the well-known contagious character of hysterical behaviour, and this is a problem that the occupational health specialist is liable to come across.

Epilepsy

Epilepsy occurs in different forms, and the crisis may be of various types, occurring at variable intervals. It may or may not be accompanied by mental deficiency or by disturbances of character resulting from awareness of the infirmity. Although epilepsy is not characterised by any features peculiar to adolescence, it may appear during this period of life which moreover seems to favour a recrudescence of the symptoms observed. In addition, the problem of whether an epileptic is fit or not to work will become acute when the time comes for the young person to receive vocational training and seek employment. Epileptic attacks may occur either as the classical major crisis (grand mal) or be limited to momentary loss of consciousness (petit mal), sometimes accompanied by sudden loss of muscular tone, or by localised muscular tremor (myoclonic petit mal); the disease may take the form of a psychomotor epilepsy, with modifications of consciousness and the appearance of visual or aural hallucinations and sensations of strangeness or a dreamy state; certain forms may be primarily psychic, while others may affect mainly the local motor functions, etc.

In some forms of epilepsy, psychic involvement is totally absent, and this requires to be stressed. In other cases there may be disorders such as slowness of mental processes, of memory, reduced concentration, etc. For his part, the patient may accept his handicap and face it courageously with the aid of such medical assistance as is possible, or on the other hand he may react negatively, displaying discouragement and depression or resorting to dissimulation or even exaggeration of the condition; in such cases, various types of abnormal behaviour and disturbances of character following the mental shock resulting from awareness of the handicap may be observed.

The epileptic attacks may be preceded by warning signs, enabling the subject to avoid dangerous situations. Although their frequency and seriousness may be greatly diminished by adequate treatment, the necessary medication may induce somnolence and adversely affect vigilance and reaction times. It should be pointed out that concentration of attention and physical or mental activity reduce the frequency of attacks; a measure of physical activity is accordingly desirable.

It is clear from these considerations that each case has its own individual characteristics and must be dealt with on its merits. There must be close co-operation between the occupational health specialist and the family doctor or specialist responsible for diagnosis and treatment, who sees the patient regularly. In these circumstances, it is difficult to lay down any general rules; however, precautions must obviously be taken in the placement of epileptic workers to avoid accident risks for themselves and others, and to prevent their being subjected to occupational or environmental conditions which might provoke or increase the frequency of attacks (for example, lights blinking or intermittent noise occurring 8-10 times per second may trigger an attack in certain kinds of epilepsy). It will be remembered that increased risks exist in work at heights, near stretches of water, with dangerous machines and in the operation of vehicles or certain mechanical equipment.

At the pre-employment medical examination, two kinds of problems arise. On the one hand, the subject may be a recognised epileptic already undergoing treatment; in this case, it will be necessary to have a precise picture of his condition and to decide on a suitable type of work. On the other hand, the doctor may be faced with a young person who claims to be normal, but who presents certain signs suggestive of epilepsy, or who is suspected of concealing a known epileptic condition. The necessary additional examinations must be made and appropriate treatment ordered by the responsible physician.

Young epileptics must be subjected to special medical supervision, the purpose of which should be to check whether the prescribed course of treatment is being followed, to study the psychological problems entailed and help the adolescent to overcome them, and to consider possible employment in the light of the condition and its evolution. The industrial physician will also have to decide whether certain members of the staff of the undertaking should be informed of the nature of the disease from which the young worker is suffering, in order to forestall concern about certain "absences" from work or to ensure that certain treatment can be given if necessary, and this is a very delicate matter; the action clearly required by the interests of the person concerned must be taken, and the rules of medical etiquette must be respected.

"Neurological" Motor Infirmities

Various neurological lesions may result in deficiencies of posture, gait or general motoricity, generally referred to as neurological motor infirmities. Depending on the case, the lesion may be located in the peripheral or central nervous system, at the medullary or cerebral level. It will be necessary to pinpoint this location and to determine whether the condition is due to a nervous lesion, to neuritis or polyneuritis, a lesion of the motor nerve cell from the spinal cord, as in poliomyelitis, a medullary syndrome, or a cerebral lesion. In the latter case, the condition will frequently be accompanied by intellectual deficiencies and disturbances of character, and occasionally by epileptiform disorders. A complete analysis of the motor functions should be made, by studying both the functional levels of the various muscle groups and the capacity to perform the acts required in everyday occupational life. The intellectual abilities and the possible mental repercussions of the handicap should be carefully studied, and in the presence of a central lesion, the nature of mental deficiencies or disturbances of character associated with the motor deficiency must be ascertained. These investigations are most frequently performed in specialised centres, but where employment is contemplated, the occupational health specialist must be informed of the results, and there must be co-operation between him and the physicians in these centres.

The core of the problem is related to vocational guidance and training. When the mental capacities permit, everything should be done to ensure that these young persons are trained at the highest possible level, including university level, and are consequently steered towards occupations in which intellectual work predominates and the physical handicap will be less important. If the disabled person is directed towards manual work, the occupational health specialist will be concerned with overcoming the problems which will inevitably arise in fitting the work to the handicapped worker.

Mental Deficiency

The evaluation of mental deficiency is not simple; it is a serious mistake to classify as a mental deficiency a mere case of immatureness or a lack not of ability but of intellectual background due to insufficient education. From a study of mental age and intellectual quotient, it is possible to distinguish between slight remediable deficiencies, medium deficiencies, serious deficiencies and irremediable deficiencies. Much caution, common sense and patience in repeating tests must be exercised, however, before classifying a young person in one or other of these categories. Sometimes a diagnostic or therapeutic centre makes insufficient allowance for the fact that the determination of the existing deficiency and the evaluation of its possible extension or improvement share the limitations of the psychometric methods used.

The human problems posed by mental deficiency are extremely important and are likely to become still more so as medical science provides effective weapons against the infectious and other diseases which formerly proved fatal to such young persons at an early age. There are, first and foremost, the problems facing the young handicapped person, who must be in a position to lead as normal and useful a life as possible, without feeling abandoned or rejected by society, a situation to which some of them can be very sensitive, since mental deficiency is often accompanied by exaggerated affectivity. Also, acute problems sometimes arise within the family, for which a feeble-minded member may represent a very heavy burden. Lastly, there is the problem to be faced by the community in fulfilling its humanitarian obligations towards these young persons.

The possibilities existing for education and training will differ in each case, and the choice of appropriate methods is a specialised matter. The possibilities of employment in an ordinary undertaking are very limited even for the less seriously handicapped, on the one hand on account of the level of ability required, and on the other hand on account of the occupational risks entailed and the possibility of sudden outbursts of abnormal behaviour. The chances are perhaps greatest in farming or gardening work. Various countries have created special centres for the mentally deficient, which sometimes take the shape of "sheltered employment" centres.

The Skin and Dermatological Problems

Bodily hygiene and personal cleanliness are mandatory, and the sooner these habits are acquired, the better. Yet this is not always so easy, when one considers that in many houses in industrial and agricultural areas there is no bathroom. Some undertakings provide shower-baths for their workers; an alternative is to use the public baths. But the essential problem is to get young workers in particular to make regular use of existing installations. Natural laziness, reluctance to follow advice and the wish to be different result in total neglect of the rules of personal hygiene by many young persons, who pride themselves on being dirty. Nevertheless, it would seem possible to change this situation by appropriate psychological action, for example, by making judicious use of the adolescent tendency to preoccupation with the self, or by presenting cleanliness as a way of being different. Good habits of personal hygiene play an essential part in helping to prevent work-related skin irritation and dermatitis.

The use of cosmetics by young girls is an increasingly frequent cause of dermatitis. It should be brought home to them that cosmetics, etc., must not be used indiscriminately and certainly not as a substitute for proper ablutions or to hide dirt. They should be made to understand that when skin irritation occurs, the application of cosmetics should be temporarily interrupted in order to avoid the possibility of contracting allergic dermatitis, and that if an allergic skin condition develops, they should immediately stop using any kind of products, to avoid polysensitisation, and should consult a specialist.

It will not always be easy to impose a prohibition on the use of cosmetics at workplaces where radioactive substances or substances containing lead are handled, or to compel girls working with machines to wear protective nets.

Nail-biting can be a problem, as many adolescents indulge in this habit. Although diagnosis is easy - when the habit is noticed - attempts at treatment are largely useless, and such adolescents should not be employed at workplaces where there is a risk of ingesting poisons by this route.

The skin is at increasing occupational risk generally, and the reported cases of irritation and of contact or allergic dermatitis are rising by leaps and bounds. Dermatitis is one of the commonest problems in occupational medicine, and adolescents appear to be particularly prone to it. While some cases of contact dermatitis retain this character, others quickly take on the form of allergic dermatitis, as in the case of a marked eczema appearing suddenly after fifteen days to three weeks of exposure to a certain hazard. Fairly frequently, however, a dermatitis which was initially of the contact type progressively assumes an allergic character.

More and more frequently, the question will arise whether a young worker with a skin condition should be removed from exposure to a skin hazard, on account of the possibility that the condition may be aggravated by prolonged exposure, and that polysensitisation may appear. A check should first be made to see whether the appropriate preventive measures are being taken (cleanliness, immediate washing after any accidental contamination, avoidance of use of irritating, abrasive or degreasing products for skin care).

It will also be useful to follow the course of the condition for some time, since fairly cften (in 15-20 per cent of cases) a resistance appears, accompanied by attenuation and sometimes by complete disappearance of the symptoms.

Eczema is one of the most frequent skin diseases in adults and also in adolescents. It is caused by an allergy triggered by one or several causes (allergens), while at the same time certain predisposing factors which may be present to a greater or lesser degree also play an important part (mild liver disorders, neurovegetative instability, nervousness and nervous instability or neurohumoral disorders).

When eczema is found at the pre-employment examination or there is a history of eczema, one should refrain from employing such young workers at workplaces where a dermatosis hazard exists. Appropriate vocational guidance should be given. In practice, however, the number of potential allergens is so great that it is very difficult to select trades or professions in which ncne of them will be encountered. The occupational health specialist must accordingly take a calculated risk, based on his knowledge of the allergens which might possibly be responsible for the particular type of eczema involved and of the role of predisposing factors, as well as of conditions at workplaces and industrial patholcgy as it affects the undertaking. He will eliminate choices involving possible contact with products known to be especially allergenic.

If eczema is found at a periodic examination, a possible occupational origin will have to be investigated. Here, the main information will be derived from questioning and from an intimate knowledge of the workplaces involved; patch tests conducted in good conditions, avoiding in particular any period during which the condition is developing, can supply valuable information. It will be necessary to decide whether the young worker should be withdrawn from exposure to the putative hazard; possible preventive measures, and the possible development of a resistance to the allergen, should in particular be taken into account.

Acne is extremely widespread; at least 10-20 per cent of adolescents suffer from this form of dermatitis, which may be considered as specific for this stage of life. It is frequently a cause of concern to them, to girls in particular. Patience and tact are required in dealing with complaints about this condition. A substantial role is played by predisposing factors, such as minor endocrine or liver disorders, the state of the neurovegetative system, fatigue, psychological disturbances and so on. Treatment is often ineffective. Persons prone to this condition are also predisposed to oil acne; it follows that at pre-employment examinations one should to the extent possible avoid referring these young persons to employment involving contact with lubricating oils and greases; on the other hand, active employment in the open air is extremely suitable.

Young persons also suffer frequently from warts. Sometimes some minor epidemics occur, and the possibility that the infection is being transmitted by water (as when several persons wash their hands in the same bucket, for example) must be considered. Excessive sweating of the hands occurs frequently during puberty, and may be very hampering for some kinds of work. Reference should also be made to hives (a fairly common and usually trivial complaint in adolescents), parakeratosis, skin infections and erythema due to circulatory stasis, which is found chiefly in girls.

Athlete's foot (dermatophytosis interdigitale) may be very widespread and even assume epidemic proportions in some undertakings. It is frequently transmitted via the wooden floorboards in shower-baths, which harbour the fungus and transmit it to the foot through pressure arising on contact. The works doctor may have to recommend suitable preventive measures (elimination of wooden floorboards, disinfection, etc.).

Mention should also be made, on account of their significance in certain areas, of parasitic diseases (lice, scabies, etc.) as well as of ringworm (tinea) infections in adolescents, which are generally a form of favus. In girls, one should be careful not to confuse small deposits due to overuse of hair spray with lice larvae (nits).

The Teeth and Dental Hygiene

Young persons between 16 and 20 years of age are especially prone to suffer from dental caries. This particular sensitivity may be related to the fact that calcification of the permanent teeth continues during adolescence. Dietary habits certainly play an important part at the end of childhood, and in the course of adolescence exposure to aggressive physical and chemical agents increases; particularly harmful factors include the abuse of sweetmeats and the consumption of acidulated drinks and very hot or very cold food and drinks. An adequate intake of fluorine, which is valuable in childhood before the appearance of the first and second teeth, does not appear to be quite so important in adolescence.

A dental examination should always be performed at the employment and periodic medical examinations. Steps should be taken to ensure that young persons receive the necessary medical treatment, but all too often, unfortunately, they go to the dentist only when they are in considerable pain; thus it will be necessary to resort to every form of persuasion in order to encourage them to have regular treatment, including emphasis on the disadvantages of bad breath in social intercourse. Appropriate health counselling should not be omitted (cleaning of teeth, suitable and unsuitable diets, etc.).

When dental treatment is undertaken too late, extraction of much-decayed teeth may be the only remedy. Moreover, in areas where there is little opportunity for dental treatment, extraction is practised on a large scale, with the result that on examination, many adolescents will be found to be lacking several or all of their teeth. Ill-fitting false teeth may then pose substantial problems.

In many developing countries, dental health is a very serious problem and possibilities of treatment are minimal. In these circumstances, poor dental hygiene may affect nutrition and general health.

Digestive System

Many adolescents suffer from various digestive disorders, which usually take the shape of functional or psychosomatic disorders, in which neurovegetative dystonia plays a substantial part, and characterised mainly by gastric pains, heartburn, dyspepsia with flatulence, and vague abdominal pains, etc. "Liverish" adolescents, who complain of difficult digestion, somnolence after meals, migraine, etc. are also fairly frequently encountered.

These digestive disorders may be caused by a lack of alimentary hygiene or an unbalanced diet. In some cases they may arise because, not being provided with the rich foodstuffs needed to meet his nutritional requirements, the adolescent compensates lack of quality by increased intake. In others they may be due to an insufficiently varied diet, with a superabundance of certain foodstuffs such as sweetmeats, pastries and sweet dishes in general, or, on the contrary, very spicy or very salty dishes, or pork meat preparations. The habit of taking meals irregularly and hastily, gulping them down, also contributes to such disorders, as does immoderate use of tobacco or alcoholic drinks. Account must also be taken of the affectivity of adolescents, of the possible existence of a state of psychological stress, and of the lability of the autonomic nervous system which is habitual at this age. Additional examinations should, of course, be made whenever they appear necessary.

A number of different organic infections may be found. Gastric or duodenal ulcers may be present, even in very serious forms, at the end of adolescence. The diagnosis of "chronic appendicitis" is highly debateable and may cloak a number of different conditions. Attention should be drawn to the problem of ulceromembranous colitis which, although not particularly frequent in adolescence, very often makes its appearance at the end of this period of life; the significant role played by psychic factors in this disease is well known, and conditions at the workplace, or problems of adaptation to work, may not be foreign to it.

A careful exploration should be made for hernia. In boys, this is typically inguinal and mainly congenital in character. Hernias of a different origin (relaxed ligaments, abdominal muscular insufficiency, etc.) occur mainly in adults. If a hernia is found at a pre-employment medical examination, this will constitute a contra-indication for a variety of occupations (heavy physical labour, work entailing thoracico-abdominal effort, load carrying, work involving standing for long periods). If the hernia is found following a violent effort, the delicate question of accidental hernia and entitlement to workmen's compensation will then arise. Adolescents with hernia should be advised to consult a surgeon and not to hesitate to undergo an operation performed in good conditions.

Sight must not be lost of the possible existence of intestinal parasitosis, which may be very prevalent in young persons in certain areas. The condition may be caused by infestation by threadworm (enterobius), roundworm (ascaris), tapeworm (taenia) or hookworm (ancylostoma) in hot and humid regions, or may have been contracted as a result of holidays in such regions or work in mines (although young persons are in principle prohibited from such work).

Genito-Urinary Apparatus

Albuminuria is very frequently found at pre-employment or subsequent medical examinations. Indeed, orthostatic albuminuria is a very common anomaly between the ages of 14 and 18 years. One must, however, always keep in mind the possibility that this condition may be a symptom of urinary infection, or of nephritis or some other serious renal complaint which must not be overlooked. Within the context of occupational medicine, certain simple additional examinations may already be performed (urinary sedimentation, albumin level, particularly in morning urine after emptying the bladder the previous night). But the other necessary examinations will also have to be made in order to gain a complete picture, for it must not be forgotten that the factors causing functional albuminuria may also cause the appearance of urinary signs pointing to latent nephritis.

In considering the functional albuminurias which have no pathological significance, a distinction is made between simple orthostatic albuminuria which occurs only on standing for long periods, orthostatic albuminuria which is related both to the upright position and to the prone position with lordosis, and albuminuria resulting from diet, exposure to cold or very heavy physical exercise, work or sporting activities. Thus, a transitory albuminuria was found in 40 per cent of apprentices who participated in a cross-country race. The existence of functional albuminuria need impose no limitation on occupational or physical activity; at most, the subject will do best to avoid work involving long periods of standing. Nevertheless, it is advisable to check up on this condition at subsequent medical examinations.

If a chronic kidney disease is diagnosed, however, the situation will be quite different. The subject must be made to follow appropriate medical treatment and must avoid tiring work performed standing, or prolonged, substantial muscular effort. Exposure to conditions conducive to infections of the respiratory tract (irritating vapours and gases, damp cold, fog and mist, etc.) must be avoided. Appropriate action in the field of vocational guidance and training or retraining will be required.

Enuresis may still occur during adolescence; the existence of a lesion of the urinary tract, or a possible psychological cause, should be investigated.

In boys, the problem of malposition of the testicles (ectopia) should always be resolved before adolescence, but unfortunately this is not always done. In the first place, testicular retraction should not be confused with ectopia; in the former, the gonad is forced back by a powerful cremasteric contraction, and this is a temporary and trivial phenomenon, whereas in the latter case, the testis is not in the bursa because it is abnormally positioned (ectopia). The subject should be examined for a concomitant hernia or an anomaly of the inguinal region, which are often associated with this condition. He should be advised to seek possible appropriate treatment.

In girls, the occupational health specialist will often be confronted with gynaecological problems, either during the pre-employment or periodic examinations or on account of repeated absenteeism or momentary interruptions of work to seek rest in the medical service round about the time of menstruation. These can be a real problem in undertakings employing large numbers of women workers. Such disorders as absent, excessive, prolonged, or painful menstruation, as well as pre-menstrual tension syndromes, leucorrhoea and vaginitis may be encountered. Whenever the diagnosis is not clear, the subject should be referred to a specialist for a detailed examination. It still happens all too often that consistently painful menstruation is neglected, until finally an ovarian cyst or endometriosis is found, for example. One should also not forget that exposure to certain poisons may result in abnormal menstruation, and also that psychological problems - which may have arisen at the workplace - may, for example, be responsible for stoppage of menstrual flow.

A venereal disease problem may exist. In some regions, syphilis is endemic or increasing; moreover, although the "pill" eliminates the fear of undesired pregnancy, it may increase the risk of venereal infections. An infected adolescent may be ashamed to consult a doctor for this reason, and may perhaps do so more easily during a preventive examination which he must take in any case. It may also happen that a pregnancy, of which the subject was unaware, can be discovered on the occasion of a periodic medical examination.

Especial attention must be drawn to the problem of the possible exposure of young workers to the hazards of ionising radiation. The doses which are significant from the standpoint of the genetic risk for the population in general are those received before or during the procreative period. It must not be forgotten that all doses, no matter how small, received by any person who will have children will have the effect of increasing the frequency of mutation. According to present knowledge, the genetic risk is proportional to the accumulated radiation dose in all such persons, and there is no threshold of safety. Young persons under 16 years of age must absolutely not be employed in occupations involving any risk of exposure to ionising radiation, and a particularly severe limit on maximum radiation doses must be imposed until the age of 18; better still, the total prohibition on exposure to ionising radiation should be extended until the age of 18, as is the case in several countries.

Endocrine System

If glycosuria is found, the possibility of juvenile diabetes should be investigated, but not, of course, before a check has been made that the substance excreted really is glucose and not something else (e.g. pharmaceutical preparations, albumin, fructose, etc.), or that the condition is not a simple renal glycosuria (diabetes innocens). Questioning will show whether the fundamental symptoms (incessant morbid hunger and thirst, increased urination, loss of weight) and a possible family history of diabetes are present. Measurement of blood sugar is indispensable, and may in some cases be performed by the industrial medical service. A precise diagnosis showing the complete clinical picture will then be required, and these are frequently done in special centres or during hospitalisation.

Juvenile diabetes differs in some respects from adult diabetes. It is usually a serious condition, sometimes accompanied by substantial loss of weight (in contrast to the obesity often found in adult diabetics). At this age, the usual degenerative ocular, renal and arterial complications are not normally found, since these only appear after a lapse of several years. The treatment must include administration of insulin. In prescribing a diet, account should be taken of the fact that the patient is a growing young person likely to display considerable physical activity, and whose nutritional requirements are correspondingly great. Juvenile diabetes is rather difficult to control consistently, and there is a risk of convulsive crises caused by blood sugar deficiency. Care must be taken to ensure that the young person is adequately informed about his condition and participates actively in its treatment, despite the difficulties which may sometimes be experienced in persuading young persons to accept the routines and restrictions required to prevent acute crises and long-term complications (frequent analysis of urine, carefully-planned diet, regular treatment and feeding, etc.).

A diabetic person with a well-controlled blood sugar balance and who takes adequate care of himself can and must lead a normal existence. Some physical activity, without exaggeration, is very desirable. There should be wide areas of employment open to young diabetics, who should not be turned away systematically, as is too often the case. Nevertheless, they should not be employed on jobs in which a crisis due to blood sugar deficiency might have serious consequences for themselves or others (operation of vehicles, earth-moving and hoisting equipment, work at heights or with other potentially dangerous machinery, etc.). They should also not be employed on work which interferes with meal timetables or otherwise exposes them to irregular feeding, or work involving a definite risk of minor injuries (on account of the complications ensuing from infection, to which diabetics are particularly exposed).

Industrial medical services will frequently be called upon to participate in the treatment of diabetic workers, who often ask, with the consent of their personal physician, for insulin injections and urine analysis to be performed there. The works doctor should verify that the treatment is being regularly followed, encourage the young diabetic to accept the constraints resulting from his condition, and collaborate with the latter's physician. At periodic examinations, special check-ups should be given to detect the first signs of any complications, having regard in particular to the special sensitivity of young diabetics to tuberculosis; it will be necessary to verify whether the job and the workplace are still appropriate, in the light of the course taken by the disease and its treatment. Account will also have to be taken of a sometimes substantial variation in sugar metabolism occurring in the course of a day, as a result of insulin injections and food intake; at times, a certain blood sugar deficiency may exist, resulting in debility, a loss of muscular strength and decreased vigilance.

Enlargement of the thyroid unaccompanied by clear signs of over- or under-activity of this gland is a fairly common finding in some persons, especially girls, at clinical examinations. Most usually, these are cases of physiological hypertrophy ("adolescent's goitre"), which may be related to the endocrine changes occurring in adolescence, and to an increased requirement of thyroid hormone resulting from the growth and maturation processes. There may be a background of inadequate iodine intake, and this possibility should be kept in mind, more particularly in areas where the diet contains little iodine. If the increase in thyroid volume is diffused, without any nodules or signs of thyroid dysfunction, the condition is most usually an insigificant anomaly which will tend to disappear with age; in such cases, the best course is simply to await developments, while perhaps ordering an increased iodine intake, rather than to embark on unnecessary endocrinotherapy, with the attendant risk of complications. Where necessary, additional examinations should be arranged to confirm the diagnosis and determine the treatment required. The possibility of hypothyroidism, hyperthyroidism, or inflammation of the thyroid gland should be considered; it can be very difficult to distinguish between simple adolescent goitre and lymphocytic thyroiditis. The presence of a nodule can never be treated lightly in an adolescent, and must always be regarded as a possible sign of incipient carcinoma.

A number of anomalies of puberal development may be encountered, which may be related to endocrine dysfunction. The systems comprising the hypothalamus, pituitary and adrenal glands and gonads may be somewhat slow in beginning to operate, with the result that puberty is retarded. A pituitary deficiency with dwarfism may be found, more frequently in boys. There may be delayed puberty accompanied by substantial growth in height; in boys, the possibility of testicular insufficiency (eunuchoidism), either as a primary condition or as a sequel of injury, orchitis, bilateral ectopia, etc., should be considered. In cases of adrenal over-activity, precocious puberty will be observed in boys, and delayed puberty, accompanied by virilism, in girls.

Many other disorders of the endocrine system may occur in young persons, but as all of them are relatively rare compared to those just referred to, it would not be rational to consider them in detail here.

CHAPTER III

THE INDUSTRIAL PHYSICIAN AND THE YOUNG WORKER

Initial Reception in the Undertaking

One's first job, one's first contract with an undertaking and one's first working day are important steps in one's life. These are things that one does not easily forget, and very often the first job will influence the course of one's subsequent career. The way in which one was set to work, first impressions, the manner in which one was received and the extent to which one succeeded in fitting into the working environment are especially important factors. For the first time, the young worker comes face to face with professional life, and has to begin to equate his personal aspirations, expectations and hopes with reality. First impressions sometimes have a crucial effect on the relationships in process of establishment between the worker and the undertaking, which is likely to affect his attitude towards his work and the manner in which he performs it. The way young workers are introduced into an undertaking is also very important on account of the difficulties that may arise through the failure of adult workers to accept them, or through the conflicts which are liable to occur.

Young persons beginning work stand in need of a considerable amount of general and practical information. This is usually supplied by the personnel department, and relates to such matters as hours of work, remuneration, holidays and formalities required in the event of absence, illness or accidents, as well as to plant, departmental and workshop rules and practices. General information on the activities pursued by the undertaking and its various departments, with special reference to those of the branch in which the young person will work, is particularly useful. Emphasis must be placed on safety and health matters, such as action to be taken in the case of technical incidents, accidents or fire, on procedures for giving the alarm in emergencies and for calling in the medical service, as well as on the available equipment (fire fighting, first aid, etc.). The occupational safety and health service and the members of the safety and health committee certainly have an important part to play in this respect. General information will also have to be given about the use of changing rooms, refectories and other facilities provided for workers, as well as the functions of the social service and the assistance that it can provide, trade union activities, the safety and health committee, and cultural or sports activities. In this way the young worker will be able to form an over-all impression of the undertaking and how he can fit into it, and will not be at a loss on finding himself unceremoniously plunged into a wholly unfamiliar situation. Furthermore, a careful initial briefing will help the adolescent to avoid mistakes due to ignorance, which might lead to disputes with other workers.

The first contact with the works medical service will take place right at the outset, in connection with the medical fitness test. A friendly reception, during which it is made clear that the service is interested in him as a person and not only in his possible diseases, abilities or disabilities, will help greatly to establish a good relationship between the medical service and the young worker, as is indeed essential if the various purposes aimed at are to be achieved. It will be necessary to explain the role and operation of the medical service and the organisation of first aid to the young worker, and to give him indispensable information about the occupational risks that he may encounter and the precautions that he should take, stressing his responsibility towards himself and towards others in helping to prevent occupational accidents and diseases, and to improve safety and health conditions in the undertaking. He must be made to realise that his co-operation is really needed. This initial reception is just as important as the medical examination proper, and the industrial physician and nurse must be aware of this fact.

The young worker's hopes must on no account be dashed in the first few days by an unfavourable reception, or by the lack of any kind of introduction, combined with total indifference. The natural anxiety which he is already experiencing must not be made worse by such a situation. Most often, the young worker will be very keen to work, to learn, to understand and to be efficient and useful, he will want to do good work and to progress. But if he is met only with indifference or perhaps even with hostility, if his experience is restricted to routine and botched jobs, all his good intentions will quickly vanish, and this new potential, on not being put to good use, will soon be swallowed up within the common herd. Depending on his character, he will be lost or hostile, and will respond with indifference or with aggressive reactions.

An adolescent needs to feel that he is accepted by his workmates and is really participating in a common effort. While he is eager to give of his best, he also wants very much to know on whose behalf he is doing so, and he must be acquainted with his chiefs, who must have the authority corresponding to their positions and responsibilities, but must also be approachable, sympathetic human beings with whom it is possible to talk freely. The young worker must be made to realise that he is being trusted, that he is being given a chance and that it is up to him to make use of it. He must know that his work is duly appreciated, if there is anything wrong he must be told, but he should also be told when what he does is right and well done. One must refrain from systematically handing him only the dullest or dirtiest chores in the beginning, and treating him as the servant of his seniors.

When the young worker arrives at the workplace, he must be shown his workstation and briefed on the nature of his work. He must be introduced to the department manager and to his immediate supervisor and workmates, as well as to the person in charge of safety and health matters. Someone must be made responsible for initiating him systematically and progressively in his work. This person must be aware of the importance of this responsibility and must convey the necessary information, reply to questions, help the young worker with technical problems, teach him to assume responsibility, direct his initiatives and help him to adapt to his work. An adolescent is in need of information, understanding and unwavering support. The reception and introduction of young workers must not be left only to the personnel department or medical service, but must first and foremost be taken in hand by the workers already employed in the undertaking, since in the last analysis, it is at the workplace itself that the success or failure of his integration into the working environment will be decided. Workers' organisations and their representatives should be made aware of their rightful mission in this field. The young worker should also be encouraged to participate in the life of the undertaking, including trade union activities, the work of the safety and health committee, and such social, cultural or sports activities as may exist.

In practice, unfortunately, young workers are rarely informed adequately about the essential aspects of their working life, they are rarely given a systematic initiation or welcomed into the undertaking as they should be. If at all, this is done mainly in large industrial undertakings and on behalf of young persons who will go straight into a fairly high post; the others are left to sink or swim alone, or with the mere aid of such advice and assistance as may be condescended from time to time. And yet, where systematic arrangements for their reception, introduction and initial processing do exist, it is found that the young workers so treated tend to feel happier in such undertakings, to change their jobs less often and to advance more quickly in their occupations.

Purposes of Medical Supervision

Exercising a proper supervision over the state of health of young workers involves doing everything necessary to prevent the appearance of health problems in the young person, and to overcome them where they do exist. It is a question not only of diagnosing or treating illness, but of maintaining the physical and mental health of each individual at the highest possible level.

When the occupational health specialist receives an adolescent, he should consider how he can be of service to his visitor. When the adolescent leaves his office, it must be in the knowledge that the visit was beneficial and that he has learnt something of which he was previously unaware. The industrial physician should be concerned with the young worker himself, and not merely with diseases, affections or anomalies present in certain adolescents. His sphere of interest must extend beyond occupational or juvenile pathology to include the adolescent as a person, his development and problems. He must always be ready to give young workers a helping hand in their evolution towards adulthood.

Young persons stand in need of special supervision for a number of reasons. Adolescence is a period of physical and mental growth and maturation characterised by perpetual change, during which a state of balance is achieved only fleetingly; the normal course of this development must not be jeopardised by unsuitable employment or working conditions.

For treatment to be really effective, and fitness for the widest possible variety of employment to be safeguarded to the utmost, it is important that any pathological conditions, factors or anomalies should be discovered as quickly as possible. This applies equally to somatic and mental or psychological affections and to lack of adaptation to work. Moreover, today's young workers are tomorrow's senior work force, the physical and mental health and therefore the working capacity and efficiency of which will be conditioned by the level of medical care provided to them in their youth.

The pre-employment medical examination must lead to the adolescent's placement in a suitable job, and prevent his being directed to employment which would be particularly hazardous for his health or development. The physical or mental efforts required by a given task must not be allowed to aggravate existing handicaps; account must also be taken of the existence of a special sensitivity to certain occupational risks.

For the young worker, the time has come to choose his occupation, he is looking for an aim in life and for opportunities of learning, he needs to have a clear picture of his physical and mental capacities, he needs advice, and the occupational health specialist, if he so wishes, can play a vital part in helping him to see things clearly. At the same time, he will have to discourage the adolescent from launching out in directions which will lead him nowhere because he lacks the indispensable physical or mental capacities.

Adolescence is a formative period, a period of education and training. If the young person is to learn a trade, he must start doing so properly right from the outset, and part of his vocational training consists in acquiring safe and reliable techniques and working habits which are in keeping with the dictates of occupational health and hygiene. The works doctor must play an active part in informing and training young workers in occupational safety and health matters.

Lastly, the pre-employment medical examination will provide data for a general assessment which will serve as a point of reference for the ensuing medical supervision. Such an initial picture is necessary for the purpose of assessing the subsequent physical, mental and occupational development of young workers, while at the same time it will serve as a useful yardstick in detecting incipient work-related disorders.

The periodic medical examinations permit the early diagnosis of the first signs of an occupational disease or of particular sensitivity to certain factors, thereby enabling steps to be taken to ward off ill health or effect a cure before any serious lesions appear. They will also show up any signs of excessive fatigue or lack of adaptation to work or working conditions, so that the necessary measures can be taken to fit the work and the conditions in which it is performed to the requirements of human physiology, or to direct the worker to employment better suited to his capacities and wishes.

The periodic medical examinations also afford an opportunity of keeping a check on the course of development and maturation processes in adolescents, and of taking any action that may seem necessary to promote them. They also may lead to the discovery of anomalies, deficiencies or diseases of recent origin, while the course of any conditions found at the pre-employment examination can be followed, and the continuing suitability of the job for the young worker can be verified.

Although the purpose of all this medical supervision is to safeguard and improve the safety and health of young workers, it is sometimes objected that the pre-employment medical examination is a selective process designed to eliminate anyone who might become a burden on the undertaking, and that the periodic examination may be used as an instrument for ordering transfers to the detriment of the workers concerned, or for putting an end to contracts of employment and terminating the worker on grounds of unfitness. It must be admitted that both of these interpretations are possible.

For example, one can decide that a young worker with asthma cannot be employed in a bakery because of the risk that his condition might become worse, but one might also decide that he is unfit in order to avoid having to compensate an occupational disease. Or one could certify as unfit for work as a mason in building construction a young person with a history of repeated respiratory infection, either because the working conditions might be an aggravating factor, or in order to avoid the cost of frequent sickness-induced absenteeism. It will be easily understood that it is in the interests of a young person with a deficiency entailing an increased risk of occupational accidents - and of his workmates - to refrain from employing him at a dangerous workplace. On the other hand, since an occupational accident can involve an undertaking in compensation costs and also in expenses due to loss of equipment and production, one might say that one is selecting only persons who are fit for the job and rejecting others because their physical or mental deficiencies might be too costly for the employer.

In the examples just given, the medical examination is exactly the same, regardless of the intentions behind it, and the resulting verdict of fitness or unfitness is also exactly the same. If the worker or his representative regards the works doctor as being "on the boss's side" his decisions may be disputed. In order to avoid such pitfalls and preclude mistaken interpretations, the industrial physician must be completely independent and must earn and retain the trust both of the workers and of management.

It is the industrial physicians who are in a position to decide whether the functions of their medical service will be confined essentially to medical selection, routine administration or health check-ups, or will be broad enough to ensure the effective protection of the health and welfare of young workers. It is for them to decide whether in their relations with young workers their role will above all be that of a physician, purely and simply. It is their responsibility to impose a proper approach to the medical supervision of young workers, in a climate of total independence.

The medical supervision of the health of young workers by works medical services must be comprised within an integrated system organised at the national level to safeguard the health of young persons and promote their development from every point of view. The industrial physician must not consider only the events taking place in the undertaking during working hours, but must not hesitate to co-operate closely with all the other bodies which are actively concerned with young persons and with their health, and in particular with the medical services responsible for examining young persons while at school, in connection with vocational guidance, during vocational training, or in relation with sports activities.

There is no doubt, moreover, that occupational health specialists, who are knowledgeable about people, who are familiar with adolescents, their peculiarities and needs, who are acquainted with the hazards of various occupations and the repercussions on the organism of exposure to harmful industrial substances and environments, and who have a good knowledge of working conditions, can play an essential part in advising the competent authorities about the action required to safeguard the health of workers, and of young workers in particular, reasonably and effectively. The information that industrial physicians, who have to face all these problems every day, can provide is of capital importance and must be taken into account both by the authorities and by emplcyers' and workers' organisations. Industrial physicians, or their professional associations, have a duty to provide complete information based on practical experience. They should participate actively in the framing of rules and regulations concerning the protection of the health of workers, all the more so since they will finally be responsible for implementing these instruments.

The Medical Examination

Medical History

An adolescent is no longer a child, whose illnesses and condition are described by the parents, nor is he yet an adult able to give a clear description of his symptoms (or a hypochondriac prone to exaggerate minor complaints). In general, an adolescent, although attaching great importance to his person, is rather reluctant to speak of health problems. If he finds a doctor who is ready to listen and to help, however, he will be more willing to supply a history, give explanations, advance opinions and put questions.

Frequently, he will stand in need of reassurance, for even normal adolescents can be concerned by imagined health problems. He must be dealt with patiently and given whatever explanations he seeks. An adolescent will not forgive an attempt to fob off his unfounded but nevertheless real anxiety with a shrug of the shoulders, or, worse still, jocularity or sarcasm; nor, for that matter, will he rest satisfied with an attitude of uncertainty, for the last thing he wants are doubts which will only exacerbate his natural anxiety.

Sometimes an adolescent may be looking for support and approval (but not for systematic approval given easily and without justification), while others, on the contrary, will be on the defensive and may display a hostile or even aggressive attitude.

Little by little, the adolescent will come to realise that the preservation of his health is no longer his parents' responsibility, but his own. He needs someone who is interested in himself, rather than in his illnesses, and an atmosphere of confidence must be established. It is absolutely indispensable that young persons should be certain of enjoying total medical secrecy. It is possible for doctors to establish much deeper and more fruitful relationships with young persons than with adults, for very often, if he is able to win their confidence, the doctor will be regarded by adolescents as possessing the key to all life's secrets. But in order to achieve such a position of trust, continued awareness of the particular psychology and problems characterising this stage of life is necessary.

The medical history should cover significant illnesses of the parents (e.g. diabetes, epilepsy, cardiac or nervous diseases, etc.) as well as of the young person (diseases of childhood, accidental injuries, other disorders, operations, vaccinations, etc.), and should include data on current habits (usual diet, physical activities, sleep, any tendency to fatigue, use of alcohol, tobacco, pharmaceutical preparations, etc.). In the interests of speed and completeness, one may use a prepared list mentioning the various points to be checked. The questioning should also cover the various body systems and functions, with a view to pinpointing pathological factors or particular sensitivities which have a bearing on fitness for work.

The subject's previous occupational history (if any) and educational history should also be covered, noting any difficulties encountered, and the extent of physical and psychological adaptation to these situations achieved. The kind of education and training received may also be recorded.

If the adolescent is already employed in the undertaking, the medical record should indicate what work he has done, for how long and in what circumstances. Note should be taken of the subject's reactions to his work, whether he likes or dislikes it and finds it easy or difficult. The expectations, ambitions and motivations of the young worker, and the extent of his integration in the working environment and in the life of the undertaking in general, should also be explored.

It is very important that sufficient information is obtained, since effective action is possible only in clear situations. Moreover, an insufficient or incomplete medical history may lead to errors of diagnosis or interpretation, or may result in a condition going undetected. During the questioning, the doctor will have an opportunity of detecting psychological disorders or signs of lack of adaptation which, if left to pursue their course, may have serious repercussions on the mental and even physical health of the subject, whereas advice or even minor action taken sufficiently early might eliminate them, as the causes are frequently insignificant, or have a positive influence by setting the young person back on the right road.

Clinical Examination

By definition, the clinical examination must be a general examination covering the various organs, systems and body functions. A complete examination should be made both before employment and at periodic check-ups.

In the case of examinations performed on adult workers to prevent the development of occupational diseases, it may be logical in some cases to look more particularly for certain symptoms. But young workers are in a different situation, and in their case it is necessary to follow step by step and in detail the essentially dynamic processes of physical, mental and psychological development and maturation, and it will be from the evolution noted at successive examinations that the most useful information can be obtained.

Moreover, in the case of adolescents, the medical examination must look beyond the prevention of disease to the confirmation of fitness or discovery of disabilities. It must be directed not merely at certain organs which are particularly sensitive to the effects of a given industrial poison, but at the whole physical and mental personality of the adolescent in all its aspects.

It must also be remembered that adolescents do not always wish to have their deficiencies discovered, and quite often the existence of a problem will be revealed only by some minor sign found at the clinical examination. Sometimes also, the fact that the doctor appears to be spending more time on such and such an organ may prompt an adolescent to speak up about his problems.

Certain additional examinations form part and parcel of the general clinical examination. This applies particularly to urine analysis, which must be treated as mandatory, and repeated at each subsequent medical check-up.

The clinical examination also includes certain biometric data, such as height and weight. In the case of adolescents, it can usefully cover all the biometric data whose measurement at successive examinations will provide a picture of growth in size and weight; also of interest is the measurement of vital capacity and its evolution in time. The data required to evaluate biological age and the evolution of puberty should also be recorded.

Since one is dealing with young workers, it will be of interest to study their muscular strength and its development, as well as physiological reactions to effort and the degree of adaptation to it. These tests can supply extremely interesting information, especially if they are repeated, and should be included in any industrial medical examination.

A test for short-sightedness, using test types, is generally made during the clinical examination, and the importance of subjecting all adolescents to systematic eyesight tests should be emphasised.

It is particularly important, also, that young workers should be given hearing tests. Useful information can already be obtained from a quick check-up by conventional acoumetric methods and this should be done systematically.

It is desirable that young persons should undergo an annual X-ray examination of the thorax, with the object in particular of preventing tuberculosis; this is treated in greater detail in the relevant section.

Cutaneous tuberculin sensitivity tests should be performed systematically on all adolescents.

Additional Examinations

On the occasion of the pre-employment medical examination, special examinations may be made to determine fitness for the job in view, either because it may involve exposure to certain occupational risks, or because particular aptitudes are required in order to perform the work. At periodic examinations, special check-ups may be made to detect the first signs of occupational poisoning or disease. Lastly, additional examinations may be called for owing to the discovery of an anomaly, an illness or a deficiency in the subject; their purpose will be to confirm the diagnosis and elucidate the extent of fitness for work.

The list of additional examinations that may be required is long; the following are some examples of those which are performed relatively frequently:

- Blood tests, including haemoglobin, erythrocyte and leucocyte levels, leucocyte formula and also, according to the case, bleeding and coagulation times, tourniquet test and thrombocyte count. These blood tests providing various analyses will be made in particular at the pre-employment medical examination, for example to assess fitness for employment in conditions involving a risk of exposure to certain toxic solvents, lead and certain other poisons or ionising radiation (young persons over 18 years of age).

- At the periodic examinations, the blood may be tested for factors denoting incipient poisoning (e.g. basophil stippling) or changes in the levels of certain constituents brought about by various poisons (e.g. cholinesterase level in exposure to certain pesticides), or again, for the presence of the poisons themselves (as in poisoning by lead or carbon monoxide, for example); it may also be explored for biological signs of possible kidney or liver disorders due to certain chemicals or their metabolites.

- In addition to the usual chemical analyses, the urinary sediment may be analysed, or tests made for metabolites the presence of which above certain levels indicates incipient poisoning (e.g. urinary coproprophyrin in workers exposed to lead, trichloroacetic acid in workers exposed to trichloroethylene), or, again, for the presence of the poisons themselves (lead, arsenic, mercury, etc.), these analyses being made as part of the regular medical supervision of workers exposed to the specific occupational risk.

- Detailed ophthalmological examinations may be performed either at the pre-employment examination, when the nature cf the work so requires (colour vision, binocular vision, insensitivity to dazzle, visual field, etc.), or at periodic examinations, to check that visual acuity is maintained or that it has not been affected by exposure to particular risks (infra-red or ionising radiation, etc.).

- The hearing of workers exposed to noise should be regularly checked by audiometric examinations in which the threshold of sound perception is checked up to higher frequencies, with the object of detecting any loss at around 4,000 Hz, indicative of occupational hearing loss. Where a hearing loss is found, its influence on speech comprehension may be studied by oral tests.

- On occasion, tests may be made to investigate the cardiac and circulatory or respiratory functions, or aptitude for effort, for the purpose either of confirming a diagnosis or of assessing the suitability of the subject for certain types of heavy work or difficult working conditions (e.g. hot work). Such tests will be found particularly helpful, for example, in selecting young adults to man rescue teams in mines; such personnel must be in excellent physical condition, since in the event of a disaster they will be called upon to make enormous efforts and to work extremely quickly, often in a very hot environment (mine fires) and while wearing respiratory protective equipment (for protection against poison gases and a lack of oxygen in the air).

- Psychotechnical tests may be used at the pre-employment examination, in order to investigate psychomotor capacity (dexterity, speed, accuracy, reaction time, etc.), for tasks requiring certain aptitudes, or various mental capacities (e.g. vigilance (concentrated or at intervals, etc.), instantaneous recall and long-term memory, comprehension of orders or code signals, etc.). The subject's reactions may also be studied by placing him in a situation similar to that with which he will be confronted at the workplace; such tests are frequently used for vehicle operators. Certain tests may also be used at periodic examinations, for example, to evaluate a state of fatigue (reaction time, fusion frequency, tests with series of colours, numbers, etc.). While some tests may be carried out at the medical service, most frequently they will be entrusted to a specialist.

The above are examples of examinations that may be desirable to assess fitness for a particular job, or mandatory in view of the special occupational risks involved. In addition, a variety of special examinations may be required, depending on the case, in order to confirm a diagnosis or evaluate the seriousness of an occupational disease, symptoms of which have been found at a medical examination, and decide on a course of treatment. These additional blood tests (or X-ray or other examinations, etc.) will usually be the responsibility of the medical services providing treatment. However, the industrial physician should be informed of the results, so that he will be able to gain a picture of the extent of the person's fitness for work and suggest a suitable job for him, on the principle that nobody should be debarred from working and that the aim should be to employ everyone, whatever his deficiency may be, in the tasks which he is capable of performing, within the limits of the existing employment opportunities in the undertaking. Although ordinarily the industrial physician does not request such additional examinations directly or perform them himself, he should at least inform the young person concerned of the situation and advise him to consult the appropriate physician or medical service without delay. He has an important part to play in persuading young workers to submit to treatment, which is not always an easy matter. In addition, he should ascertain that they have actually undergone the necessary examinations and are following the prescribed course of treatment or heeding the advice that has been given to them. Collaboration between industrial and general physicians and health services should be much closer than is sometimes the case.

Recording Results: The Medical File

The industrial medical service must keep a medical file for each young worker, providing a continuing picture of his health and physical and mental development, as well as an over-all picture of the general health of the young workers in the undertaking.

ILO Recommendation No. 112 concerning occupational health services in places of employment provides that these services should open a confidential personal medical file for every worker at the time of the worker's pre-employment examination or first visit to the service, and should keep the file up to date at each succeeding examination or visit. Such files should contain information about the worker's medical history, the clinical examination and any additional or special examinations undergone, as well as the verdict reached as to fitness for the proposed employment. The medical record or file should be so designed as to facilitate the entry by the physician of all important information, for example by including a list of points for checking or consideration. The establishment of a model medical file or record could with advantage be considered by a panel composed of all the occupational health specialists in a given area, or by the relevant professional associations; such an approach would allow a measure of standardisation and facilitate the exchange of information between regional or national occupational health services. Sometimes a model medical record is required by law, and in such cases, the occupational health specialist should be consulted without fail when it is drawn up.

Certificates, reports or other similar documents should be transmitted to the young worker himself, to his parents or guardian, to his employer and, as the case may be, to the official service responsible for supervising the implementation of the rules and regulations relating to the medical examination of young workers. In drawing up these documents, medical secrecy and etiquette must be strictly observed, and this also applies to the establishment and keeping of the medical record by the occupational health service. Information given in a medical certificate to be transmitted to the employer should never contain anything of a confidential character such as the diagnosis of congenital deficiencies or affections found at the medical examination. Particular precautions must also be taken regarding the data inserted in the health record. The young person himself or his parents must be given such information as it is medically justified to provide. The form in which information must be supplied to the competent official services must be carefully chosen so that medical secrecy is fully preserved.

Young workers should keep some written evidence of the examinations that they have undergone. Certain national laws and regulations provide that every young worker must have a personal health card or medical booklet in which certain information obtained at medical examinations is entered. In theory, the idea of such a personal record which would follow an individual from birth onwards throughout his life appears interesting, as it provides a link between health supervision exercised in early childhood, during the school years and in adolescence. Moreover, such a personal record could also help to establish a link between the works doctor and the young worker's personal physician, and to avoid duplication, with particular reference to the useless repetition of radiological examinations, with the attendant risks resulting from the accumulated radiation doses received. But the existence of an official document (even if it can be communicated only through doctors) containing medical information (even if it is kept vague and diagnoses are omitted) is often viewed as a source of unacceptable constraints, and in practice such systems are often not very efficient. Accordingly, if he should come across any pathological findings, the industrial physician will do best to hand the adolescent a letter addressed to his personal physician.

The occupational health specialist should inform the parents or guardians of the young person of his findings concerning the latter's fitness for work and state of health; if necessary, he will recommend that the latter should consult his doctor or go to the dentist. When the circumstances make this desirable, he should also remain in regular contact with the family doctor or other doctor looking after the young person. It is interesting to note that the regulations in force in the Federal Republic of Germany provide for a medical record in three parts, of which one remains on file, the other is sent to the parents or guardian and the third is the medical certificate sent to the employer. These three parts contain appropriate information, which differs according to their nature; however, ingenious use of carbon paper enables all three documents to be grouped in a single composite form and typed up simultaneously.

The employer, thus, should be informed of the results of the medical examination (fitness or unfitness, temporary fitness, contra-indication of certain kinds of employment) by means of an appropriate document. The international standards, and most of the national laws and regulations, relating to the pre-employment medical examination of children and adolescents provide that these may be admitted or maintained in employment only on presentation of a medical certificate of fitness for the work entailed. This fitness can also be attested to by an observation on the work permit or work book. The employer should retain these documents on file and keep them at the disposal of the competent authorities responsible for supervising the application of the rules and regulations relating to the protection of young workers.

It may be considered useful for the administration responsible for enforcing the application of the rules and regulations on the supervision of the health of young workers, as well as the public health authorities, to be kept informed of the results of these medical examinations by appropriate documents. This data will enable them to keep a check on the state of health of young workers and young persons in general throughout the country, and will give them a clearer picture of the action required to protect the health of the latter effectively. National legislation may call for the reporting of cases of occupational diseases, or cases of compensable occupational disease, or of all cases of affections relating in any way to work.

ILO Recommendation No. 112 also provides that the occupational health services should compile and periodically review statistics concerning health conditions in the undertaking. They should also maintain appropriate records so that they can provide any necessary information concerning the work of the service and the general state of health of the workers.

Frequently, the employer will ask the occupational health service to submit periodic and annual reports, as other services are required to do. The works safety and health committee, if any, and the workers' representatives should also be kept informed of the work done by the occupational health service and of the health problems encountered. National laws and regulations sometimes require the occupational health services to submit a detailed annual report, drawn up according to a certain model, to the responsible administration.

Conclusions of Medical Examination: "Fit" or "Unfit"?

The medical examination of a worker may lead to an evaluation of his state of health, of his general fitness for work, or of his fitness for a particular job. It is necessary clearly to understand the difference between these three kinds of evaluations.

The assessment of general state of health is a medical opinion which concerns the doctor, the person examined and his parents. If any important deficiences or lesions are found, there may be grounds for a verdict of infirmity.

The assessment of general fitness for work corresponds to an evaluation of the young person's possibility of employment on the general and local employment market, taking into consideration his state of health. If he is suffering from any kind of affection, deficiency or anomaly, a degree of disability may be fixed which will determine the decrease in the adolescent's scope for employment compared to the jobs offered to a healthy adolescent of the same age and level of training, living in the same region in the same circumstances. This, therefore, is an entirely relative concept and it is accordingly understandable that, depending on the circumstances and the physician, the percentage of disability regarded as corresponding to a given health deficiency may differ substantially.

The medical examination given to young workers, however, is essentially designed to test fitness for particular kinds of work. In submitting his opinion to the employer, the industrial physician does not need to pass judgment on the young worker's state of health or evaluate his fitness for employment on the open market, possibly qualifying this by a percentage of disability, but to come to a conclusion concerning the young worker's fitness for a given range of jobs performed in a given range of circumstances, or for employment in a particular job.

In the former case, the young worker's aptitude to perform certain operations in certain occupational circumstances will be determined; for example in some cases heavy work may be contra-indicated, but work performed bending over will not, or again, exposure to bad weather may be contra-indicated, but not work calling for sustained vigilance, and so on. Lists have been drawn up of the categories of work and kinds of occupational circumstances with respect to which it is particularly useful to obtain an opinion.

In the latter case, the question to be answered is whether a given adolescent is fit for employment at a certain workstation. In answering it, of course, a knowledge of the job and of the circumstances in which it will be performed is required.

If the industrial physician confines himself to the former, or more general, case, somebody will still have to decide whether or not a particular adolescent with particular contra-indications may be assigned to a specific job involving particular requirements, and different interpretations are possible. This problem does not arise if the industrial physician himself also decides on the latter or specific issue. In making out medical certificates, one should never resort to the easy way out by inserting such vague remarks as "in good health" or "normal" (with reference to the state of health), or, again, "fit" (indicating that the working capacity of the subject is "normal"). On the contrary, an assessment of fitness or unfitness should always be accompanied by an indication of the particular job and workstation to which it applies.

If an adolescent has a certain deficiency, entailing a certain decrease in his working capacity and hence in his scope for employment, this will often be because he is (totally) unfit for some kinds of work, whereas on the other hand he is (totally) fit for other kinds of work. If, then, he is assigned to the latter, he will be indistinguishable from a normal individual, and it will be as though, from his own point of view and from that of society, his deficiency did not exist. The industrial physician must be inspired primarily by this aim. Rather than merely brand an individual as being "unfit for such and such a job", the medical certificate should indicate that he is "fit for such and such another job".

When the situation is less clear-cut, i.e. when a person is not wholly fit for a given job but not wholly unfit for it either, the industrial physician may adopt an intermediate solution, for example by making employment subject to certain conditions or to special supervision, and by granting a certificate of fitness only for a limited period. The international standards, moreover, make room for such provisos. He may also call for additional examinations in order further to clarify the person's state of health and facilitate his decision.

In very many cases, the young worker will be found to be in good health, with a sturdy constitution exempt from anomalies, deficiences or illness, and there need then be no hesitation in declaring him fit for normal employment for his age. But in other cases, when particular findings are made at a pre-employment or periodic examination, the question of fitness or otherwise will arise, and answering it will often prove to be no easy matter.

The issue is affected by a whole series of factors, for all of which due allowance must be made. On the one hand, there is the nature and seriousness of the affection or deficiency, and the possibility that it may become progressively worse, either spontaneously or under the influence of the work performed or the circumstances surrounding it. On the other hand, there is the astonishing capacity for compensating their shortcomings displayed by many adolescents, particularly when they are strongly motivated. Last but not least, one must always remember that one is not dealing with a mere "medical case" but with a young man or girl who must as far as possible be able to live the same life as other young persons in their age group, who must be able to work, to earn a living and to fit into society.

Thus, the industrial physician is often faced with the need to take a very delicate decision:

"Fit for Work"? But then, is he not laying himself open to criticism for failing to do everything in his power to safeguard the health of the young worker? The employer may complain that he may be obliged to hire or retain young persons who are unsuited to the work involved, while for their part, the workers' representatives might consider that the physician is not doing his duty, which is to see that the health of young workers is protected. If things go badly, the young person or his parents may even turn against him.

"Unfit for work"? If so, at least no one will be able to accuse the industrial physician of lack of foresight and he will be fully covered; but may this not also be "an easy way out", which makes insufficient allowance for the fact that he is dealing not with a disease but with a person who has a right to work and to have a normal social life? It would not be desirable for a medical verdict delivered with due regard to personal and other circumstances to be replaced by the compulsory application of lists of disabilities entailing automatic disqualification for certain jobs. Such lists may be useful, but only as guides, and on condition that they are not compulsory and cannot be used to call into question the responsibility of the physician, the employer or the worker himself; otherwise, they would perforce have to be applied stricty, and the scope for employing handicapped persons would be greatly limited, to the detriment of many workers with anomalies or deficiencies.

The verdict of fitness or unfitness must take into account the personality and social circumstances of the subject, as well as his state of health. Armed with adequate information on all three counts, the occupational health specialist can appreciate whether or not any risk implicit for the young person in the proposed employment is worth taking or not. Since he alone is fully conversant with the subject's state of health, he is the only person able to decide whether the subject should be offered the job for which he is applying or not. But since as a doctor he cannot reveal the medical aspects of the situation, he is automatically debarred from disclosing all the grounds motivating his decision. The latter is accordingly not open to discussion and no justification for it may be given to anyone whatsoever, except possibly another physician, within the limits of medical etiquette. Clearly, therefore, it is of vital importance that industrial physicians should be fully independent of both the employer and the workers, and this independence is the best guarantee that the real interests of young workers will be safeguarded.

Disputes, if any, arising out of this decision should be settled at the medical level. For example, if a young worker or his parents disagree with a ruling of unfitness, they should ask their family doctor to contact the industrial physician; very often, the problem will thereupon be cleared up, but if the two physicians are unable to agree, medical arbitration may then be resorted to, and this procedure is in fact provided for in the legislation of some countries.

If the workers' representatives consider that a ruling of unfitness made by the industrial physician has harmed one of their workmates unduly, by causing him to be transferred or to lose his job, they should ask a doctor whom they trust to contact the industrial physician, whereupon, again, the situation will frequently be cleared up. Also, if the trade unions consider that the workers examined should be given fuller information about their health, or that the unions should be informed in greater detail about the health of the staff, they should ask a doctor attached to them or with whom they are in touch to contact the industrial physician and go into these questions with him. Medical problems should be dealt with according to the rules of medical etiquette, and the right of every worker to medical secrecy must be respected; such problems must remain within the medical field and must not go beyond it.

Particular and often delicate questions of fitness arise when it has to be decided whether or not initial signs of injury to health resulting from work, or of an occupational disease, found at a periodic examination justify withdrawing the worker from the occupational hazard to which he was exposed. There is little objective information on this subject, and opinions often differ as to the action required when the first very minor signs of occupational disease are found. Should the worker be removed from the relevant exposure and from his work? When should this be done, and for how long? And when will he be in a position to resume his work involving exposure to this occupational hazard? While such questions are already difficult to answer owing to the lack of relevant scientific data, they become even more delicate when one takes into account the wishes of the worker, who frequently wants to keep his job instead of being given other, perhaps less paid work or even dismissed on the grounds that all the jobs that he could do involve some degree of exposure to the same hazardous condition. The verdict of fitness or unfitness is perhaps particularly important for young workers, as it is delivered at the start of their occupational life, and a negative decision may ruin their hopes of practising their chosen trade, and thus affect them very greatly.

Advising Young Workers

A young person seeking employment must be given all relevant information at the pre-employment examination about any occupational risks that he may encounter at his workplace, and must also receive precise, clear and complete information about the health precautions that he should take from the outset in order to avoid falling a victim to occupational diseases. This information must be repeated and added to as required on the occasion of his subsequent visits to the medical service. In doing this, occupational hazards must be placed in their proper perspective, stressing the fact that work well done is work done cleanly, and that the tendency of certain older workers deliberately to disregard health precautions does not speak in favour of their experience or qualifications - quite on the contrary! One must also explain to him why certain working methods which are taught, such as manual load-handling techniques, are important for the preservation of health.

If the pre-employment medical examination, and, to an even greater extent, the periodic examination, are just that and nothing else, young workers will soon come to regard them as useless, and the effectiveness of the medical assistance which it is desired to give them will suffer accordingly. These medical examinations should serve as a starting point for the provision of information, advice, guidance, training and education. The usefulness of occupational health will be fully recognised, and the occupational health service will be able to play its part to the full, only if this condition is effectively met.

Young workers should be given a picture of their physical and mental capacities, and when a deficiency is found, they should be informed of it to the extent that is medically justified, emphasising the scope for and advantages to be derived from treatment and indicating the possibilities that exist of overcoming or compensating the deficiency. Sometimes general health counselling will be sufficient (more attention tc balanced nutrition, participation in certain sports, longer rest periods, use of footwear providing better support, how to avoid chills, eat, smoke less or if possible not at all, etc.). At other times, the advice given will be related to occupational life (use of means of personal protection, hand-washing immediately after accidental contact with irritating substances, avoidance of solvents for hand-washing, need for training in correct load-carrying techniques, etc.). Lastly, the occupational health specialist may advise the adolescent to see a dentist or to consult a physician or medical service where he can receive the treatment corresponding to his physical or mental condition.

When the deficiency is so marked as to entail a reduction in working capacity, appropriate measures must be taken for the resettlement or physical or occupational retraining of the person concerned. There must be co-operation towards this end between the occupational health service and the employment, educational and social services, and effective contact must be maintained between these services until the young worker has been reintegrated in the workforce. ILO Recommendation No. 79 concerning the Medical Examination for Fitness for Employment of Children and Young Persons indicates the measures that should be taken for young persons found by medical examination to be unfit or only partially fit for employment. Such young persons should receive proper medical treatment for removing or alleviating their handicap or limitation; they should be encouraged to return to school or guided towards suitable occupations likely to be agreeable to them and within their capacity, opportunities of training for such occupations being provided to them, and they should have the advantage of financial aid, if necessary, during the period of medical treatment, schooling or vocational training.

In addition, the medical service should stimulate awareness of occupational health problems in young workers. The necessary explanations should be given to enable them to understand the importance from the medical point of view of improvements in environmental conditions of workplaces (lighting, noise abatement, fresh air supply, removal of gas and fumes by exhaust ventilation, etc.) and of technical action aimed at fitting the work to the realities of human physiology, or at protecting the worker. The industrial physician must help to promote a spirit of safety-mindedness in the undertaking, and must participate actively in training and education given with this aim in view.

The role of the occupational health service, the purpose of the periodic medical examinations and the organisation of emergency medical care and first aid must also be explained. The young worker must know that the works doctor is there to help him, and that he can consult him about any questions connected with health, or indeed about any of his problems.

Supervision of Health Conditions at Workplaces

The international standards adopted by the ILO, and many national laws and regulations, specify that young persons may not be admitted to employment unless they have been found fit for the work on which they are to be employed by a thorough medical examination. It is accordingly important that the occupational health specialist should not merely be armed with theoretical knowledge, but should know that the young person before him is scheduled for employment at a given workplace, in a given location and in particular circumstances, with all of which the physician should be familiar.

Furthermore, it is only by gaining a thorough knowledge of workplaces and of the circumstances and conditions in which work is performed that the physician will be able really to grasp the nature of the problems arising, and to supervise the health of young workers most effectively. The finding of certain anomalies at medical examinations will draw his attention to certain questions that merit investigation during a tour of the workplaces, while vice-versa, certain observations made at workplaces will lead him to check that there have been no adverse consequences for the health of the workers. A critical comparison of the results of medical examinations and of workplace inspections may also lead to changes in the layout of workstations and improvements in the organisation of work, applying ergonomic principles and having regard to the fact that young workers are employed there.

The supervision of conditions at workplaces includes the study of environmental factors such as temperature, humidity, air currents, radiant heat, vapours and gases, dust, lighting, noise, vibration and odours.

It is generally considered that omfortable ranges of temperature and air humidity are the same for adolescents as for adults, and that standards laid down for the latter also apply to the former. Some national laws and regulations lay down maximum values for ambient temperatures of workplaces, depending on the work performed there. A great deal of research has been done in this field, and a series of sometimes complex graphs and simplified tables or formulae are available for defining ranges of comfortable temperatures.

Although young persons generally see better than adults in poor light, are better at distinguishing details and require less time for observing objects, this does not mean that adolescents do not need such strong lighting as adults. On the contrary, at workplaces where young persons are employed, particular care should be paid to the quality and intensity of lighting, in order to avoid visual fatigue, and the deterioration of eyesight which might ultimately ensue. Standards of lighting prescribed for adults also hold good for young workers; they are sometimes determined by national laws and regulations and sometimes by bodies concerned with occupational safety in health, in co-operation with associations of lighting specialists.

As has already been pointed out, there are clear indications that adolescents, particularly in the 14-16 year age group, are particularly sensitive to the risk of hearing damage resulting from exposure to high-pitched noise. They also appear to be particularly sensitive to the effects of such noise on the cardiac and circulatory system and the autonomic nervous system. Standards or graphs have been established indicating maximum tolerable noise levels at various frequencies for continuous exposure during eight hours daily. Some national laws and regulations lay down maximum tolerable levels, or levels of habitual exposure to noise, in excess of which exposed workers should receive regular audiometric check-ups.

Odours are also a factor in the workplace environment. They may be characteristic of a workshop or workstation (rubber, textile, leather and plastic industries) or be very frequently and generally encountered (hot oil, trichloroethylene, formaldehyde compounds, etc.). They may give warning of a technical defect (e.g. mechanical breakdown, inadequate ventilation, leaking duct). Workers frequently attach great importance to odours; although their significance as warning signs of technical incidents must not be overlooked, it is also necessary to emphasise that there is no relation between the smell given off by a substance and its possible toxicity. The smell of benzene, for example, is not at all marked and more agreeable than that of trichloroethylene, which is less poisonous; again, chlorine gives off a strong smell even in weak concentrations, whereas carbon monoxide is odourless.

In the case of many irritating and poisonous gases, as well as of the dusts of certain mineral or vegetable substances, threshold limit values or maximum allowable concentrations of these substances in the atmosphere of workplaces have been laid down for conditions of continuous exposure during a working day of eight hours; they are sometimes determined by bodies concerned with occupational health and sometimes imposed by national laws and regulations. These values may be determined on the strength of different criteria, a fact which partly explains the differences that are to be observed between the figures given; rather than speak of a limit value, it would in any case be better to accept the existence of a limiting zone. Moreover, their interpretation gives rise to great difficulties, since a worker is only very rarely exposed continuously to the same concentration of a harmful substance in the course of his working day. The values indicated should be regarded as a first stage which must necessarily be reached in the fight to control environmental pollution at workplaces. For certain substances, e.g. carbon monoxide, graphs are available indicating the maximum allowable concentrations in relation to duration of exposure. Generally speaking, different threshold limit values are not set for young workers and for adults; nevertheless, young workers are prohibited by national laws and regulations from employment on certain kinds of work or exposure to certain occupational hazards.

The choice of a suitable place of work is a delicate question which is sometimes hard to solve in practice. If the pace is too quick, the physical or mental effort required is too great and cannot be maintained for long periods or with sufficient safety. The optimum pace, providing the highest output from the standpoint of both quality and quantity, never corresponds to the fastest rate at which a person can work. On the other hand, the pace must not be set too slow, or a kind of lethargic condition may ensue, which may in its turn induce symptoms of fatigue.

Moreover, while a highly irregular working pace, interspersed with periods of inactivity due to a poor organisation of work, has an adverse effect on productivity, repetitive work performed at too regular a pace engenders a sort of monotony which may also have adverse repercussions on the quality and even the quantity of output. Account must likewise be taken of differences between individual workers, which also have their importance. While some workers need to work at a steady, not too rapid pace, others are capable of working at very high speed during relatively short periods, separated by moments of relative rest. The need to keep pace with the machine may force a worker to slow down his natural working pace, or to exceed it. These problems may be particularly important for young workers, who are not accustomed to working at a set pace and may experience some difficulty in adapting themselves to this requirement.

The presence of the industrial physician at the workplace is a token of his interest in the workers and their work, which adolescents in particular appreciate greatly. The fact that the doctor took the trouble to study the existing problems on the spot cannot fail to have an extremely good effect on the relations of young workers with the occupational health service. In particular, they will usually be more inclined to accept any suggestions that are made to them and to acknowledge the need to follow any advice given when they see that these observations are the outcome of investigations on the spot and are not prompted merely by theoretical considerations.

The discreet and effective supervision that the occupational health service is in a position to exercise over health conditions in and the state of upkeep of premises and equipment used for the welfare of workers is also likely to have a positive effect on the training and general education of young workers, and on their psychological attitude towards the undertaking and towards work as a whole. Clean and tidy washrooms, conveniences and cloakrooms can only encourage young workers to adopt good habits which will serve them well both at the workplace and in private life. A particular aspect of this general supervision of health conditions, which is usually the responsibility of the occupational health services, is the supervision by the works doctor of plant refectories, their premises, equipment and staff, as well as of the composition and quality of the meals served there. As a result of current trends towards the establishment of large industrial centres, accompanied, in the event of decentralisation, by the increasingly long distances between the workplace and the home or living quarters, young workers are more and more frequently obliged to take their meals at the factory or other undertaking. In such circumstances, the inspection by the works doctor of the composition of meals to be served to adolescents and young persons is especially important, for two main reasons, namely, on the one hand, to ascertain that the rations served ensure an adequate, balanced and nourishing intake of food and, on the other hand, to induce young workers to follow the advice in dietary matters that he will have had occasion to give at medical examinations, i.e. to pursue their dietary education - which is all too often deficient - at a practical level and inculcate healthy feeding habits in them.

Occupational Accidents

The problem of occupational accidents is first and foremost a problem of industrial safety, and the industrial physician is consequently somewhat prone to leave it to the specialists, whereas in fact he has an essential part to play in this field.

At pre-employment and periodic examinations, he is in a position to discover causes of lack of adaptation to work which, depending on developments, may in time become causes of occupational accidents. He can recognise predisposing factors such as fatigue or stress, and deal with them before they become dangerous. In particular, he should keep a regular check on the health of workers in safety-sensitive jobs, in which an individual weakness may place an entire group at risk.

In inspecting workplaces, he may find control stations and panels with dials which are hard to read, poorly laid out or badly lighted, with poorly designed or inaccessible controls. He may find that the environmental noise level hampers the clear receipt of orders and instructions, that the general lighting is insufficient, or that other unfavourable environmental factors are adversely affecting occupational safety. Cleanliness and tidiness are twins. While cleanliness is important from the standpoint of health, it is no less so from the standpoint of safety, as it applies to flooring, for example; similarly, an untidy workshop, with tools and equipment lying around and getting in the way, is a frequent cause of occupational accidents.

Furthermore, the physician is the only person to be informed, thanks to the dispensary records, of the occurrence of a number of minor accidents which were not otherwise reported; however, a minor accident may point to the existence of an extremely dangerous situation, in which the injured person was lucky to get off lightly. From his examination of the injuries resulting from occupational accidents, the physician will derive certain clues as to how the accident took place. It follows that he is in a good position to know the hazards existing in the undertaking.

Lastly, the adverse consequences of occupational accidents may be minimised by efficient arrangements for emergency and follow-up treatment. The health service can play an important part in this respect by keeping first-aid kits fully stocked, ensuring that enough stretchers are available and training members of first-aid and rescue teams. Time spent on occupational safety is not wasted. By devoting an hour to safety matters, the physician may save several hours of consultations and treatment later on. His discussions with engineers, the few extra minutes taken to give advice to young workers at periodic medical examinations, the memoranda to management and the other miscellaneous chores which apparently have only a remote connection with his medical duties may save him the onus of having subsequently to perform work which is definitely of a medical kind but the nature and circumstances of which he can only deplore ... and the young workers concerned as well.

As his technical knowledge is limited, the industrial physician may feel less well placed than the engineer to deal with safety questions, although his contribution can be very effective for several reasons. Technicians who live in a state of permanent co-existence with dangerous situations end up by being largely blind to them, whereas the works doctor, coming fresh to a situation, will see them. Many accident causes are very simple, and once pinpointed, a moment's thought will suffice to find a remedy: obvious examples are untidy workplaces, unguarded machinery, slippery floors, the lack of guard rails or failure to use personal protective equipment. Moreover, many safety problems may be very complicated when they are tackled on a technical level and relatively simple if a less conventional approach is adcpted; how often, indeed, a safety device evokes the reflection: "it's simple, but someone had to think of it". Not infrequently, also, there is general awareness that a dangerous situation exists, but nothing is done to remedy it because the relevant information is not being fed through to management, so that the pertinent comments made by the workers do not reach the ears of the person having power to change the situation. The industrial physician, however, being attached to the management, is in a position to make the necessary observations at the proper level to ensure that they are acted on.

In most economic sectors of most countries, there is general agreement that the accident rate for young workers is higher than that for the workforce as a whole. Also, it is often found that young workers tend to have more serious accidents than their seniors, although, here, much depends on the workplace and circumstances and on the extent of the restrictions placed on the employment of young persons.

The fact that young workers are especially prone to industrial accidents is one reason for taking an active interest in this question, but there are others. It is particularly regrettable when the victim of an accident entailing a disability is a young person, because the handicap will stay with him all his life and may perhaps prevent him from doing not only the work on which he had originally set his mind but many other things that he would like to do, as well participating in the sporting and social activities which are normal at his age.

Several factors may be invoked to explain why young workers are more accident prone, and they should be taken into account in making the necessary arrangements for safety measures.

First of all, there is the fact that an accident is generally a chance occurrence due to the existence of a dangerous situation which has nothing to do with chance. The question can be studied by considering in turn the dangerous situation, the factors tending to transform a dangerous situation into an incident or accident, and the aggravating factors tending to turn an incident or minor accident into a more serious accident.

When a dangerous situation has been noted, a first step towards preventing an accident has already been taken; when it has been realised that the situation is dangerous, a second step has been taken; and when a timely decision has been taken to remedy the situation, an accident has been avoided. But the manner cf working, too, may also multiply the rate of appearance of dangerous situations in a substantial measure, and this must be avoided.

A young person starting work does not always perceive danger correctly, or else he has an impression of insecurity, of vague danger, but does not always clearly perceive the relationship between an eventual situation and the idea of danger or a particular hazard, or comes to the wrong conclusions. While the pouring of molten metal will impress him as a highly dangerous operation, he will use a grindstone without wearing goggles and will not look where he is putting his fingers. This lack of experience is particularly evident in mines, where, for example, creaking noises or small falls of stones may be meaningless, or may mean that the roof is settling or about to cave in and that one should get clear of the area, and only the old-timers can interpret these signs correctly. Young workers, therefore, stand in need of information which cannot be found in books and training which cannot be received only at school but must be given at the workplace, and mainly by older workmates, who have a definite responsibility in this respect which they should not neglect and of which the workers' representatives should be aware. The lack of perception of hazards by young workers due to ignorance must not be replaced merely by a disregard bred of habit.

A young worker has not yet mastered the most efficient, fastest and safest working techniques, his movements are not yet "professional" and automatic, and his efforts are not always well controlled and in proportion with the desired result. His eye is not yet trained, he cannot yet make reliable evaluations and "guesstimations", and he may misjudge or over-estimate his capabilities. He is inclined to use make-shift tools, rather than spend time looking for the proper ones. He lacks the experience to judge a situation correctly, anticipate what will happen and decide quickly what is to be done. These various shortcomings have the effect of multiplying the number of dangerous situations in which a young worker may find himself. From this point of view, a lot of training is required; while this is done initially at the vocational training school or during apprenticeship, it must be effectively pursued at the workplace. Towards this end, first of all, a due awareness of this need must be aroused, and secondly, someone must be made responsible for giving this training.

Even when he realises that a hazard exists, the young worker not infrequently takes unnecessary and sometimes stupid risks. A headstrong nature, a certain liking for taking risks for their own sake, a need to express or impose himself, over-keenness or an inconsistant approach may lead him to indulge in the most imprudent behaviour. For example, although he know that it is dangerous to pass under a suspended load, he thinks that if one walks quickly there is no danger and that "in any case, that sort of thing never really happens". Here too, education is necessary to give the young worker a clear understanding of his responsibilities towards himself and towards others. Not only must he be given a sense of responsibility, but he must also be given an opportunity of shouldering the responsibilities that he is capable of assuming at the particular stage of development that he has reached.

Young workers sometimes react very negatively to systematic prohibitions, whereas they are much more amenable to occupational discipline founded on a proper evaluation of hazards, provided that one takes the trouble to explain the reasons justifying it and its importance for the efficient operation of the undertaking. The explanation should cover the different safety rules and warning signs and signals, etc., and the young worker should be made to understand why the employer is not only empowered but also obliged to impose such a discipline, and require compliance with safety rules. Young workers should also be given an opportunity of participating in studies of safety problems in the undertaking;

their suggestions and opinions should be taken into account and they should be allowed to participate in the work of the safety and health committee.

Lastly, when faced with a series of events which will inevitably result in an accident if nothing is done to stop them, the young worker may not know how to do so. He may not know where to find the fire extinguisher, or the safety cutout on a machine, or he may attempt to dilute sulphuric acid escaping from a vat with water. Or again, when an accident has taken place, he may not know that the current should be cut off before rescuing a workmate from electrocution, or that he should wear respiratory protective equipment in going to help a workmate overcome by a poison gas in a confined space ... and one ends up with two victims instead of one. Here again, adequate training is clearly essential. In his ignorance, a young worker may neglect to have a small wound attended to by the medical service and the physician will see it only when it has become infected; or he may neglect to have a foreign body removed from his eye, so that a minor incident turns into a serious accident.

There are thus a number of reasons why young workers are predisposed to occupational accidents, including lack of experience, a disregard of danger, the need for self-expression and independence, etc., and while some of them are related to the worker's psychological characteristics and even, it might be said, to emotional factors, others are closely linked to questions of information and training.

It goes without saying that training is particularly important, as bad habits are quickly acquired and hard to correct. Right from the very outset, young workers must be taught correct working procedures, they must be made to understand that work well done is work done safely, and that tools and machinery, as well as electricity or radiation sources, must be treated with due care and respect. Young workers must be encouraged to take an interest in their work and to appreciate the fine points of their trade, for occupational safety is enhanced if a worker takes an interest in his work and enjoys doing it.

Special care must also be taken not to expose young persons to risks or hazardous situations for which they are not prepared. It may be necessary to debar them from a number of jobs which would involve them in personal danger or in heavy responsibilities towards other workers, either because they lack the necessary knowledge or are insufficiently accustomed to such work, or because there is a risk of their being insufficiently aware of the responsibilities entailed or of losing control if an awkward situation arose.

Although it is necessary to prohibit young persons from engaging in certain particularly dangerous work, variations between individuals must also be taken into consideration. There are young workers of 16 or 17 years of age who work much more carefully, conscientiously and safely than many adult workers. One should also not forget that a young worker to whom one entrusts certain responsibilities usually makes a point of not betraying the confidence that has been placed in him.

The establishment of lists of jobs from which young workers are debarred for their own protection must not prevent their learning the indispensable skills and receiving the training required in order to fit them adequately to perform such work in due course. In this respect, the recommendation of the European Economic Community concerning limitations on the employment of young persons provides that action should be taken to encourage the timely initiation of young workers who may be called upon to perform work regarded as dangerous for young persons, when they have reached the minimum age for such employment. This initiation should be provided in the course of vocational training or apprenticeship and should in particular include instruction in the necessary occupational safety and health precautions.

Naturally, the other aspects of occupational safety and health, such as technical safety precautions, safety devices and equipment, organisation and planning of work, choice of appropriate working methods and speeds and safety campaigns also play an important part in preventing occupational accidents to young workers as well as to their seniors.

Occupational Diseases

In considering the problem of occupational accidents and safety, regard has been had mainly to the action of agents capable of inflicting injuries and to the risk of receiving wounds, burns or injuries in general. Problems relating to the prevention of accidental poisoning must also be considered and efforts to prevent acute poisoning lead naturally to the prevention of chronic poisoning and, more generally, of all kinds of occupational diseases and work-related affections.

It is inadvisable to attempt to establish a clear limit between the prevention of occupational accidents and diseases. The problem of occupational poisoning and diseases should not be regarded as falling solely within the province of industrial medicine, any more than the problem of occupational accidents should be treated as an exclusively technical matter. Indeed, the prevention of occupational diseases is not only a medical question, it is above all a technical issue. "Medical prevention" in the shape of periodic medical examinations is not really prevention at all, since the industrial physician will only be able tc take note of health injuries which have already been suffered. He may be able to detect an occupational disease in the early stages, before it becomes serious or the accompanying lesions become irreversable, but he is not in a position to prevent its occurrence.

The question of whether young workers are more liable than adult workers to contract occupational diseases is very difficult to answer.

It is possible that a young worker may be more exposed to an occupational hazard than an adult worker in the same trade. For example, apprentice mechanics are frequently asked to clean workpieces with the aid of solvents; similarly, apprentice carpenters are often asked to apply wood impregnation products and to do small painting jobs (priming coats) and to handle various kinds of adhesives, so that they are exposed to a variety of poisonous substances, with the nature of which they are frequently unfamiliar.

On the other hand, a large number of laws and regulations are in force prohibiting the employment of young persons on work involving possible exposure to various kinds of occupational risks. In many cases, therefore, the conditions of work of young workers and of adult workers are not the same.

Young workers may be predisposed to occupational diseases by the same circumstances which predispose them to occupational accidents, i.e. psychological factors (carelessness, possibly inspired by the desire to take risks or for self-expression, etc.) and factors bound up with information and training (ignorance, lack of experience or practice, etc.). His inexperience may expose a young worker to additional risks, compared to an adult worker in the same conditions. He may not know how to weld workpieces while avoiding the noxious fumes which are produced, or how to perform certain other jobs without unnecessarily producing great quantities of dust. Moreover, young workers tend to be very active and to expend a lot of physical energy, not always in the most effective manner; this means that they tend to breathe more frequently and more deeply than adults, and they thus take in larger quantities of dust or vapours through the airways.

Information, education and training play an especially important part in the prevention of occupational hazards, and this point is stressed in many international instruments. For example, ILO Recommendation No. 117 of 1962 concerning Vocational Training provides that the training curriculum should cover not only the work skills and knowledge, but also health and safety factors involved in the occupation concerned. The importance of these problems can hardly be over-emphasised, for all too often one finds that workers, and young workers in particular, are unaware of the exact nature of the substances that they handle and of the risks involved and precautions required, either because the need to give this information has been overlooked, or because the information has been withheld in order to avoid alarming the users - an attitude which is unjustifiable and dangerous.

The question of whether adolescents are particularly sensitive to the effects of physical, chemical or biological agents present at the workplace is a highly complex one on which little research has been done, and the opinions usually quoted appear to be founded on impressions and common sense rather than on solid clinical or experimental findings.

For a given ratio of dose intake to body weight, the toxic effects may be more marked in young persons, when the disintoxication processes are less developed, but on the contrary, they may be less marked if, for example, kidney function and elimination processes are better in a young person than in an adult.

It is logical to consider that a special sensitivity exists to poisons which particularly affect the growing tissues, the organs which are especially active during growth (thyroid, etc.), the biological processes, or certain systems which display a marked instability during adolescence. It may also be considered that resistance to poisons is lowered by the increased energy expenditure required for growth.

References are made in the literature to various occupational risks to which young persons appear to be particularly sensitive and to which it is generally agreed that growing adolescents should not be exposed. Ionising radiations may be particularly harmful for adolescents, since their injurious effects on growing tissues are well known. Nevertheless, while some authors put forward this view and consider that the blood-forming system of young persons is especially susceptible, others point out that actual proof of these effects is lacking. The thyroid gland appears to be particularly sensitive to the effects of radioactive iodine. Needless to say, there is general agreement on the vital importance of the genetic risks which may accompany the exposure of adolescents tc ionising radiations. As regards benzene, variations exist in individual sensitivity to acute or chronic poisoning, apparently explained, amongst other factors, by variations in the extent of fatty tissues. Some authors consider that women are generally more sensitive than men. Some writers also claim that young persons are especially sensitive, while others find no significant differences between young persons and adults. With regard to lead poisoning, again, while some authors consider that a particular susceptibility exists during the period of active growth, other writers find no appreciable differences. Young persons may be more prone to serious and rapidly developing forms of poisoning, with neurological involvement. With respect to carbon disulphide, although the existence of a particular sensitivity has not been demonstrated, exposure of young workers to this poisoning hazard is undesirable. Experiments have shown that acute ozone or nitrogen dioxide poisoning is induced in young mice or guinea-pigs by much weaker concentrations, administered during much shorter periods, than in adult animals, and some authors consider that young persons are also particularly sensitive to respiratory irritants.

It must therefore be acknowledged that despite the conventional opinion that young persons are more sensitive to the poisonous or irritating effects of substances, the experimental and epidemiological proof that such a predisposition exists is very slight. Nevertheless, pending practical confirmation of these theoretical considerations relating to the protection of young persons, it is no doubt advisable to proceed cautiously and to limit the exposure of young workers to such substances as far as possible.

Immunisation

Good immunological protection from early childhood cnwards is essential. By the time adolescence is reached, this should long have been an accomplished fact, and only occasional revaccination should be necessary.

While immunisation is primarily a public health problem the industrial physician is also directly concerned by it, in view of the existence at workplaces of risks of infection against which immunisation provides effective protection. Occupational health services may also be called upon to participate in the general protection of the population by the periodic administration of anti-poliomyelitis vaccine or by participating in revaccination campaigns to ward off smallpox epidemics; in such cases, it is necessary to act quickly and to muster all available resources, and by making use of the occupational health services a large section of the population can be quickly reached.

Vaccination against smallpox is compulsory from a very early age - generally before the child is one year old - in many countries. Nevertheless, adolescents are sometimes encountered who were not vaccinated during childhood for various medical reasons. In such cases a problem will arise at the pre-employment examination if the young person is to be employed in work involving a possible contamination hazard, or for the performance of which vaccination is required by law, namely, is this young worker fit for such work or not? It is logical to consider that as long as the young person has not been given the necessary immunisation, he is not fit for such employment. The delicate question of whether it is justified to undertake a first vaccination at his age will then arise, as well as the question of what advice to give him in this respect, in view of the possible risks. In the case of young persons assigned to workplaces where there is an evident risk of contamination (hospitals, laundries, reprocessing of old textiles, etc.), the industrial physician must ensure that periodic revaccination is performed as required by the relevant legislation. It may possibly be performed by the industrial physician himself, who will then ascertain, by an appropriate medical examination, that revaccination is not contra-indicated. If there is a contra-indication, the question then arises of whether the young person is still fit for employment at the particular workplace and can remain there. This is a very delicate and often controversial issue.

Despite all the campaigns in favour of early and complete immunisation against diptheria, tetanus and whooping cough, many adolescents are still not vaccinated against tetanus as a matter of course. At the pre-employment examination, it will therefore be necessary to inquire whether such vaccination and revaccination has taken place. The young person may be able to produce his vaccination card. Some occupations are fraught with a particular tetanus hazard (leather tanneries, rubbish and sewage collection and disposal, market gardening, agriculture, etc.) and for certain jobs, vaccination and revaccination may be required by law. At periodic medical examinations, the industrial physician must verify that periodic revaccination has been performed in line with these provisions, or at reasonable intervals. He may possibly undertake that task himself, in which case, again, he must ascertain by examination that there are no contra-indications.

First aid to workers injured in occupational accidents is one of the responsibilities of the physician, who in addition will frequently also have to administer follow-up treatment. He will thus be confronted time and again with the problem of preventing tetanus, as a part of emergency treatment. It is a well-known fact that there is no close relationship between the seriousness of an injury and the risk of tetanus, save of course in the case of substantial wounds contaminated by dirt, with widespread tissue damage. Tetanus may equally well be transmitted via a mere scratch received from a thorn, nail, barbed wire, etc., or a haematoma under a fingernail. For this reason, some physicians consider that anti-tetanus vaccination should be given for any injury, no matter how slight. If the person has been vaccinated, there is usually no problem, and it will be sufficient to give a booster of anatoxin. But if the person has never been vaccinated, there is the question of serum prophylaxis with all the attendant present and future drawbacks. In these circumstances, and bearing in mind the large number of minor injuries occurring at workplaces (every worker receives one at least once during his working life), one may wonder whether it would not be extremely desirable for all workers to be vaccinated against tetanus. The reply appears to be in the affirmative, the more so since it is considered that the entire population should be so vaccinated. It follows that action should be taken to encourage as much as possible the vaccination and revaccination, always on a voluntary basis, of all workers, and particularly young workers. The vaccinations could be done by the industrial physician, as is already the practice in some large undertakings. This is another example of the scope for and importance of co-ordinated action in all areas of preventive medicine, for the purpose of achieving the highest possible level of public health.

A relatively high tuberculosis hazard exists at certain workplaces (hospitals, bacteriological laboratories, etc.), and if the tuberculin reactions are negative, vaccination or revaccination against tuberculosis are desirable, and are sometimes required by law. They may be performed by the industrial physician, who should ensure that continued immunity to tuberculosis is periodically checked and rechecked.

Prevention of Tuberculosis

The problem of the detection of tuberculosis may arise where a risk of occupational contamination exists (hospitals, sanatoria, bacteriological laboratories, etc.), but it also exists generally with respect to all adolescents undergoing pre-employment and periodic medical examinations. From this aspect, these examinations impinge on the general health protection of young persons, which must not be neglected, while always remembering that this is by no means their only aim.

Reference has already been made to the desirability of undertaking X-ray examinations of the thorax at the pre-employment examination and subsequently at yearly intervals at periodic examinations, not only where a particular hazard exists, but for all young workers less than 21 years old.

This examination should preferably be done by radiography, as in that way a document is available for subsequent consultation in following the course of a condition or effecting statistical studies of health problems of young persons in a given undertaking or area, while avoiding the unnecessary repetition of radiological examinations. Radiography is indispensable in the case of persons exposed to a pneumoconiosis hazard, and gives excellent information about the cardiovascular outline. Radiophotography also has a number of advantages, as it also provides a permanent record ; but whereas this technique is very suitable for tuberculosis detection, caution must be exercised in interpreting pictures of pneumoconiosis and in cardiovascular conditions. Both these examinations, when performed with modern equipment, have the advantage of involving relatively small radiation exposure. If one considers the large number of examinations made and the fact that young workers are concerned, it is obvious that recourse to radiography and radiophotography instead of radioscopy will automatically entail a substantial decrease in the dose of radiation, with its genetic hazards, received by the population for medical reasons. The importance of this aspect should be duly recognised. Radioscopy should not be used for finding cases of disease, but only as a complementary examination performed to locate an anomaly or pulmonary lesion more precisely or to study cardiopulmonary function.

Tests of skin sensitivity to tuberculin should be performed systematically on all adolescents, in accordance with established standards, and repeated periodically in order to detect any positive reaction. At present, tuberculosis prevention is sometimes based on the tuberculin sensitivity test, radiological examination of the thorax being performed only when the reaction is positive or in doubt.

Persons with negative tuberculin reactions should be vaccinated against tuberculosis when it is desirable for them to be immunised wholly or partly against this disease, either because they may be exposed occupationally to infection or possibly because they show signs of particular sensitivity. There may be advantages in having all young persons with negative reactions vaccinated, but this is intrinsically a public health problem, and there are arguments for and against such an approach. In the case of young persons who have been vaccinated against tuberculosis, tuberculin reactions should be checked periodically and they should be revaccinated if necessary, if the reactions are negative.

When a case of tuberculosis is found, in addition to ensuring that the invalid receives adequate treatment, it must not be forgotten that the case may have to be declared to the competent authorities, and that an epidemiological investigation must be carried out to discover the source of infection and find out whether the invalid has not infected other persons, or whether other persons with whom he is in contact have not been infected by the same source. This investigation must be concentrated on the invalid's relatives and friends, and in isolated areas, where the industrial physician is responsible for the treatment of workers, he will have to discharge this important duty. A similar investigation may also have to be performed among the invalid's workmates, and for this the industrial physician will, of course, be directly responsible.

Alcoholism and Abuse of Drugs

It is usually during adolescence that a person is first introduced to alcohol, tobacco and other poisonous or pharmaceutical preparations used by adults. Increasing participation in adult life, social, cultural and environmental factors, the desire for self-assertion, the curiosity which is natural at this age and mutual emulation have a significant bearing on the circumstances in which an adolescent may adopt these habits. Although this is a general problem of youth protection which goes far beyond the field of occupational health, it also has grave repercussions at the workplace, and in that respect cannot fail to be of concern to the industrial physician.

Alcohol and Alcoholic Drinks

The consumption of hard liquor or alcoholic drinks at workplaces must be prohibited. Care must be taken to ensure that sufficient healthy beverages are available to workers. In any event, good, fresh drinking water should be available to all workers at or near their workplaces, in a suitably hygienic form (disposable drinking cups, drinking fountains, etc.). Other beverages may also be provided, and the industrial physician should be consulted as to their choice. For example, it is not desirable that persons performing hot work should be provided only with iced or aerated drinks, or drinks containing stimulants, which they would consume in large quantities during the working day.

A vigorous campaign must be waged against the idea that wine or beer habitually consumed are not alcoholic drinks, or that these drinks have a favourable influence on physical strength and intellectual performance. On the contrary, stress must be placed on their adverse affects on vigilance, accuracy and quantity of work, as well as on occupational safety, on account of the increased accident risks that their use entails. It must be made clear that alcohol merely produces a purely subjective impression of improved physical or intellectual performance, whereas in reality, to an objective observer, it is obvious that the contrary is true. Workers must understand that even the smallest quantities of alcohol cause an appreciable deterioration in psychomotor and mental capacities. Everything possible must be done to counteract the psychological and social factors which can induce young workers to begin using hard liquor or alcoholic drinks (e.g. imitation, encouragement by older work ma tes, or sarcasm). The way a young worker is introduced into and received in an undertaking can play a significant role in this respect.

It is worth noting, however, that for some years there has been an appreciable reduction in the consumption of alcohclic drinks of all kinds by young persons in general, particularly, it might be said, in countries where alcoholism posed the most serious problem.

Use of Tobacco

Mutual encouragement and the wish to show proof of his entry into the adult world, or sometimes sheer boredom, or anxiety about forthcoming examinations, are some of the main reasons which may lead a young person to take up smoking. Family habits also play a significant part; if the father or the mother smokes, the son or the daughter respectively will tend to do so. Social usage, as well as the powerful conditioning due to widespread advertising, also have a considerable bearing on the frequency of smoking, the age of initiation and the tendency to progress from occasional to habitual smoking.

In examining young persons, the industrial physician may encounter all the disorders for which the use of tobacco is entirely or partly responsible, including smokers' cough, bronchitis, gastric pains, digestive disorders and various neurovegetative symptoms (palpitations, shortness of breath, mental fatigue, headache, etc.). He should tender the necessary advice and may contribute to prevention in this field.

Use of Stimulating, Tranquillising and Other Drugs

The use of amphetamine and other stimulants and "pep-pills" is very much on the increase among young persons. The problem is now encountered not only among university students preparing for examinations, but also in technical and secondary schools and even at workplaces. It also affects the developing countries, where some young persons regard these substances as "magic potions" which will enable them to do the impossible.

Physical or nervous fatigue is no longer accepted as a warning sign that rest is needed. Rather than attempt to overcome fatigue when this is necessary, the young person simply prefers to suppress it with the help of a drug. The results in practice are usually catastrophic, because the use of these substances is not kept within reasonable limits. Subjective output increases, whereas real output decreases; abuse may reach such a pitch that natural recovery is no longer possible and medical treatment becomes necessary. Sometimes, indeed, a state of dependency is reached in which any kind of activity becomes difficult without the aid of these products. Moreover, owing to the drawbacks attendant on their use, some young persons may simultaneously have recourse to tranquillisers or sleeping pills, so that a state of total unbalance is reached.

The industrial physician may come across states of fatigue or nervous depression in certain adolescents, particularly in apprentices, which are due to these causes. It should also be observed that although excessive consumption of coffee is to be deprecated, this beverage at least has the advantage of not generally entailing any serious risk that the limits of tolerance may be grossly exceeded.

It seems that a growing number of young persons lack the necessary will power and and determination to face up to their problems and cope with the tensions and anxieties of modern life. Instead, they prefer to rely on medication as a universal panacea, and start using tranquillisers. In addition, in today's highly organised society, any departure from standard behaviour is immediately regarded as abnormal, vexatious or pathological; a lively child is regarded as nervous, and a somewhat noisy and aggressive adolescent is considered to be going through a crisis and taken to the doctor, who prescribes tranquillisers for him, instead of allaying the parents' fear. A few years ago, addiction to tranquillisers was restricted largely to certain women who lacked an occupation or were depressed; nowadays, it extends practically to the entire working population, not omitting the younger element. Thus, the use of these substances is in process of invading the workplace, and although the consequences for occupational safety and health have as yet hardly been studied, the problem may prove to be a serious one.

Once he has begun taking such medication for reasons which are often futile, the adolescent only too easily tends to continue on account of the sensations that they produce, after which the next step is the use of drugs properly speaking.

Drug addicts pass through successive stages comparable to the stages of a disease. Contamination is followed by an initial phase, the development of the condition being characterised by the search for new sensations and for ever more powerful drugs. Addiction is transmissible, and is unique in that the addict is emotionally attached to his disease and does his best to spread it to others. Drug addiction has its epidemiology, its symptomatology, its mental and physical sequelae, and, sometimes, its fatal outcome. It is increasing to the extent of becoming a serious problem in certain countries. Although addiction to hallucinating drugs is at present most prevalent in student circles, it is spreading so quickly that the occupational health specialist must also be prepared to concern himself with it.

Fatigue

When the limits of physiological resistance are reached - when the processes of physical "wear and tear" exceed the body's capacity to "repair" itself - a person experiences a physical "warning sign", namely, fatigue.

A distinction may be made between local muscular fatigue, "cardiorespiratory" fatigue, "sensory fatigue" (visual or aural fatigue), and "general fatigue".

Local Muscular Fatigue

Static or dynamic effort exerted by local groups of muscles results in local muscular fatigue, giving rise to local sensations of pain ranging from stiffness to muscular cramp. Dynamometric tests show a decrease in contractive strength. The sensitivity of the muscle or tendon receptors is also disturbed, so that movements and the positioning of limbs become less precise. The reflexes may also be affected, with a raised threshold of patellar reflex sensitivity, and disturbed postural reflexes, with more marked oscillations when in the upright position.

While muscular exercise induces fatigue, with stiffness in the muscles which have performed it, local muscular fatigue appears more quickly during static effort (grasping or holding an object, etc.), including the efforts involved in remaining for long periods in an upright, bent or squatting position. Long periods spent sitting also lead to local fatigue accompanied, for example, by muscular pains in the nape of the neck and shoulders.

The onset of fatigue may be delayed by training, and adolescents have been found to be particularly capable of acquiring good muscular resistance. Moreover, whereas any person who works experiences fatigue in due course, persons in sedentary occupations become more susceptible to it.

"Cardiorespiratory" Fatigue

When intense efforts are made by a large number of muscles, energy expenditure is high and oxygen transfer must be corresponding rapid. Thus increased demands are placed on the respiratory, cardiac and circulatory functions. Since lung volumes are smaller in young workers than in adults, the former will have to breathe more rapidly than the latter in making the same effort, requiring the same oxygen consumption; moreover, since systolic output is lower in the former, the heart has to beat more rapidly in order to convey the same amount of oxygen through the circulatory system. It follows that for a given effort the cardiac and respiratory systems are stressed more highly in adolescents than in adults.

When an effort is made, lung ventilation increases and there is an increase (usually moderate) in heart rate and blood pressure. If the effort is too severe or too prolonged, "cardiorespiratory fatigue" appears, with panting or even marked shortness of breath (dyspnoea), pains in the thorax and stitches in the side, very rapid heart rate and low blood pressure. Acute cardiorespiratory fatigue may occur in performing a severe effort without preparation, the limiting factor being the oxygen deficit and the presence of lactic acid in the blood. Sub-acute cardiorespiratory fatigue may occur when the intensity of an effort is suddenly increased during normal work which is arduous but well withstood.

When cardiorespiratory fatigue occurs, effort must be interrupted for a breathing spell. In young workers, this occurs more frequently, or for less intense efforts, than in adults. It is more liable to occur in hot environments or environments polluted by irritant gases or vapours or by small concentrations of carbon monoxide. Regard must be had to these facts and to the importance of progressive adaptation to arduous work in advising young workers and fitting their work to them.

Sensory Fatigue

In work making consistent demands on the capacity of convergence and accommodation of the eyesight (office work, inspection of small components, fine soldering and electrical assembly work, etc.), or in work involving the systematic and repeated examination of objects or components (quality control, mechanical sorting operations, etc.) visual fatigue may occur. This is experienced, particularly at the end of the working day, as a feeling of heaviness in the head and eyes, accompanied by a general sensation of fatigue or sometimes by more or less pronounced headache and lachrymation. Visual fatigue can be detected by certain tests, such as the fusion frequency test. Symptoms such as feelings of heavy-headedness, headache, burning of the eyes, etc., should not be dismissed as meaningless, but their possible cause should be investigated. A defect of eyesight may be present, with defective refraction or accommodation. Or the trouble may lie in deficient lighting of the workplace (lighting too weak or too strong and dazzling, or flickering due to faulty fluorescent lighting, etc.). Sometimes several causes may be involved; for example, the use of artificial lighting with a spectrum displaying peaks which cause visual fatigue in certain persons suffering from slight refraction deficiencies.

When exposure to noise reaches a certain level, a hearing loss may be experienced at the end of a working day, which can be observed audiometrically as a lack of acuity at frequencies corresponding to the noises to which the person has been exposed. In the following hours this deficiency progressively disappears, normal hearing being re-established after an interval, the length of which depend on the degree of exposure, but which often amounts to between twelve and forty-eight hours, spent in a quieter environment. These two characteristics enable a distinction to be made audiographically between "aural fatigue" and occupational hearing loss (loss of acuity at frequencies in the region of 4,000 Hz, with no improvement following cessation of exposure).

Cases of more or less severe aural fatigue, requiring longer or shorter periods of recuperation, will be found, depending on the extent and character of noise exposure. There may be acute fatigue, sometimes tantamount to a definite sound-induced injury, or sub-acute or chronic fatigue, entailing a predisposition to occupational hearing loss. Marked aural fatigue is experienced as a decrease in acuity of hearing, accompanied by subjective buzzing and ringing noises in the ear. There may also be headaches, general fatigue, nervousness and, possibly, neurovegetative disorders.

General Fatigue

General fatigue, usually referred to as "tiredness", is a complex phenomenon which has no single cause, but a variety of causes producing the same effect; this phenomenon is intrinsically subjective, resulting not from a single sensation but from the integration of various sensations which are perceived differently in different circumstances and by different individuals. In the last analysis, one may say that "a person who says he is tired, is tired", and indeed it is very difficult to define fatigue otherwise while remaining within the conventional framework of medical, experimental and causal concepts.

While the state of general fatigue is subject to objective modifications, it appears to result mainly from the more or less vague perception of the fact that the physiological balance has been disturbed and that rest will be required in order to re-establish it. It also depends on the extent of stimulation of the centres controlling alertness by the afferent ducts or by cortical reverberation. Thus the feeling of fatigue may disappear completely following a strong stimulation or emotion, whereas monotony, and the consequent lack of stimulation, will favour its appearance.

Fatigue is a normal phenomenon, and acts as a warning sign designed to prevent the appearance of more serious disorders. The dangers inherent in attempts to overcome fatigue by doping, for example, are well known. However, fatigue may become abnormal or pathological, in particular when it is too marked to be overcome by normal rest, or occurs without any apparent reason. In time a state of chronic fatigue may be reached.

Fatigue is experienced as a general sensation of discomfort, accompanied by a wide variety of symptoms such as asthenia, somnolence (or on the contrary, insomnia), irritability, general muscular pains or cramps, various unspecific digestive disorders, cardiac over-excitability, feelings of heaviness in the legs and sometimes panting at the slightest effort. There is a tendency to decreased alertness, slowing down and errors of perception, and a weakening of the faculties of judgment and of rapid decision, as well as a drop in physical and mental performance in general. A characteristic effect is the appearance of "mental blocks" which may be revealed, for example, by the colour identification test. Working efficiency decreases, the number of errors increases, and sooner or later an incident or accident occurs.

Fatigue is perhaps one of the conditions most complained of by adolescents, and at first sight it seems surprising that young persons should be afflicted by this condition, which seems more likely to affect elderly persons or adults who are very active professionally and must also discharge their family responsibilities and cope with their personal and financial problems. However, if one considers the nature and variety of the processes taking place in young persons at this stage of life, it is easier to understand that they sometimes have good reason for feeling very tired.

Somatic development involves an extra expenditure of energy, and the digestive processes must work harder in order to assimilate the substantial amount of food which must be taken in. Thus, even at rest, energy expenditure is higher in an adolescent than in an adult. So it is not surprising that the growth processes are often accompanied by a feeling of lassitude. Sometimes the calorie intake may be limited by inadequate or unbalanced nutrition, or deliberately, particularly in young girls who are concerned about their figure, and this may lead to fatigue or to increased fatigability.

While intellectual development, general education and vocational training, a taste for rationalising, long discussions, introspection, affective hypersensitivity, day-dreaming and all the mental processes which help to shape the personality are of course indispensable, they also result in the expenditure of "mental and emotional energy" which must be regained in some suitable way. If recovery does not take place, fatigue will occur more easily, and a condition of overstress, due to school work for example, may set in.

The edification of an adolescent's social personality, with all the adaptations, conflicts, oppositions and anxieties that it entails, is also a source of mental fatigue. While an adult has family worries, an adolescent, in the process of acquiring his independence, also runs into problems with his family. An adult may suffer from work-related headache, but an adolescent may do so too, as a result of difficulties bound up with his apprenticeship, with the understanding and performance of orders, or with the demands placed on him by his chiefs.

Adolescents often indulge spontaneously in great physical activity, which is moreover indispensable for their development, but is also a source of increased energy expenditure. Sports and other spare-time activities, evening outings and social events, while being necessary for physical, emotional and social development and maturation, are some of the main causes of fatigue in adolescents.

Other causes are to be found, as in the case of adults, in the level of physical effort and mental alertness required by their work, the duration and other conditions of work (timetable, place of work, etc.) and environmental factors (heat, lighting, noise, air pollution, etc.).

In adolescents, perhaps even more than in adults, fatigue is a relative concept. Arousal of interest, motivation or violent emotion frequently cause early signs of fatigue to disappear almost magically, whereas the lack of these factors, as well as monotony, indifference, boredom, frustration or dissatisfaction with the work done on the contrary frequency lead to chronic fatigue. How many young persons laze around apathetically in armchairs, half asleep, at home, whereas on the football field or in the dance hall they are full of life and exuberance. Account should be taken of this phenomenon in the occupational field, by giving young workers special affective training, and stimulating and consolidating their interest in their work by helping them to progress and giving them worthy ambitions to aim at, instead of leaving them to remain idle or confined to cleaning and housekeeping chores in a corner of the workshop, amidst general indifference.

The additional sources of fatigue represented by commuting or by the combination of manual and intellectual work in apprenticeship, as well as by evening study and homework, must likewise not be forgotten.

When an adolescent is found to be in a state of fatigue at a medical examination, a check should be made to see whether there is no organic infection which might be responsible for it, with particular reference to infectious diseases (influenza, post-influenza syndrome, hepatitis, particularly hepatitis without jaundice, tuberculosis), to disorders of the blood (anaemia, etc.), to cardiac or endocrine disorders (hyper- or hypothyroidism, etc.), or Hodgkin's disease, etc. A check should also be made of the adolescent's diet, to ascertain whether there have been any restrictions possibly accompanied by loss of weight, and whether there is not some imbalance, especially in the form of protein, vitamin or iron deficiency. Fatigue may also be linked to incipient occupational poisoning (for example by carbon monoxide or solvents), and the absence of any such possibility should be verified.

Although the evaluation of general fatigue is a complicated and interesting problem, what matters is not the objective findings, but what the subject experiences. There is no over-all test, but only specific tests designed to probe a particular aspect of this complex condition. In order to gain an accurate general picture, therefore, a variety of tests would have to be performed within the framework of a systematic study which would hardly be compatible with the usual activities of an occupational health service. Nevertheless, it is very desirable that industrial physicians should take an interest in this problem, even if it is limited to certain aspects. Simple and rapid tests exist which can provide useful data, as long as one does not over-interpret them or attempt to turn them into specific tests for all aspects of fatigue. These include mental tests, sensorimotor or psychomotor tests such as studies of reaction time, tests designed to check the accuracy and co-ordination of movements, tapping tests, etc. Studies can be made of the performance of various muscle groups, using the ergograph, of image fusion frequency and of the appearance of delayed occular accommodation. Biological analyses can also be made (e.g. lactic acid and glucose levels in blood, adrenalin and noradrenalin metabolite levels, glucocorticoid and glucocorticoid metabolite levels, Donaggio's urine test, etc.). The clinical examination may show a high heart rate at rest, with a very rapid increase on effort, a low pressure at rest with hypotension on effort, and changes in the reflexes with either a lowering or a raising of the threshold of response.

But what matters most is to find out why a young person experiences fatigue. The aim of the medical examination should not be merely to verify that the fatigue is not being caused by some organic disease, leading to the conclusion that: "This adolescent is not ill, therefore he must be well, and his fatigue is without significance and will disappear in time, provided he has some rest." One must go further and determine the real cause of the fatigue noted, with particular reference to the factors just described. Is there a psychological basis? What are the young person's problems? Are these problems related in any way to his work? What action should be taken, and what advice should be given to him?

Although rest is the first need that springs to mind, it is far from being the only possible form of treatment. In fact rest, which leaves the adolescent to face his problems in idleness and solitude, may be useless or make matters worse. On the contrary, the young person will frequently have to be helped to solve his problems and conflicts by giving him advice and guidance, by evaluating his work and allotting interesting activities to him, and by seeing that he has sufficient physical activities with a good diet and adequate fresh air. He may have to be taught to relax and unwind and to take advantage of weekend breaks. Such education in a healthy way of life is also one of the best ways of preventing industrial fatigue. If the industrial physician only studies a few aspects of the problem of fatigue, the root cause will frequently escape him. Works doctors can no longer afford to disregard this important and widespread problem, but must encourage young workers to talk to them about themselves, their experiences, complaints, hopes and wishes; very often, in the course of such conversations, the key to the problem will emerge and effective assistance will become possible.

Sometimes nervous fatigue becomes clearly pathological, and leads to depression or even neurasthenia, characterised by great fatigability which cannot be overcome, a feeling of general discomfort, pronounced irritability, inability to concentrate, psychosomatic disorders and especially headaches, with a feeling of "pressure in the head"; it must not be forgotten that in adolescence, there is a not inconsiderable risk of further deterioration ending in severe depression with suicidal tendencies.

Ergonomics

The problem of fitting the work allotted to young persons to their capacities, taking into account their particular requirements at that stage of life, has evoked considerable attention for a long time. Laws and regulations concerning the age of admission to employment, limitations on overtime and the performance of certain kinds of work, and rest periods, make up a set of provisions designed to place reasonable restrictions on the "workload" of adolescents; this is one way of fitting the work to young persons, by providing for a transitional period between adolescent and adult life, so as to ensure that adaptation is not prevented by brutal changes, and that development and maturation processes are not jeopardised.

When an adolescent is taught to use tools properly, efficiently and safely, this should be - and usually is - tantamount to teaching him working methods and techniques which are appropriate to the desired goal, produce maximum effects with a minimum of effort and take human physiological characteristics duly into account; vocational training may thus enable a person to learn ergonomic methods of work.

The workplaces of adolescents, like those of adult workers, should be designed having due regard to ergonomic principles (e.g. rational lighting, well laid out and easily read dials and gauges, well designed control stations with suitable handles and pedals conveniently positioned and conforming to standard ergonomic models). Workstations may sometimes have to be adapted to the physical characteristics of young workers (height, weight and capacity for effort), as well as to their psychological characteristics (place of work, diversification of activities, etc.). These considerations have already been dealt with. It must not be forgotten that adolescents must above all have an opportunity of developing their faculties from all aspects and of improving their capacities and knowledge in all fields. In fitting work to young workers, the prime consideration must not be to ensure that it is suitable for the situation that the adolescent finds himself in at a given moment, but that it is progressively adapted to the progress that he must make. Every day should bring with it new knowledge, new responsibilities and new efforts requiring a higher level of skill and competence, but the rate of progress must, of course, be matched to the physical, intellectual and psychological capacities of the young worker concerned, so that it may take place as naturally and harmoniously as possible and confer maximum benefits.

This is a dynamic view of ergonomics which impinges on three fields: education (from the standpoint of acquisition of knowledge), training (from the standpoint of physical work) and psychological development (from the standpoint of progressive responsibility and integration in a working team, and their role in forming the subject's "social" personality). Thus one could say, perhaps somewhat paradoxically, that as applied to young workers, ergonomics must consist in the continuous adaptation of their work to provide for the harmonious and uninterrupted development of the capacity that young persons possess in a very high degree of fitting themselves to it.

Health Care

Industrial medicine is one of the branches of preventive medicine which, together with curative medicine, constitutes a whole entity, the purpose of which is to safeguard and restore the health of every member of the entire population. In practice, the limits of industrial medicine are variable, depending on the circumstances, on local conditions and on the extent of the available medical facilities as a whole. Why should occupational health services dispense personal treatment in countries where everyone can avail himself of high-grade personal medical services, with a free choice of a physician or hospital and with every guarantee as to discretion? On the other hand, why should they be restricted to prevention in countries where the medical needs are immense? Again, why should they dabble in public health in countries with competent public health services? Yet, on the contrary, where no such services exist, is it not the duty of the industrial physician also to see to the supply of pure drinking water and clean and healthy housing facilities, and to wage war against parasitic infection, on behalf of the community of workers for whose health he is responsible?

The first health services to be established in industrial undertakings were mainly responsible for the care of workers injured in occupational accidents. But quite soon, since social insurance and the attendant facilities were as yet virtually non-existent, treatment centres were often added, sometimes even including hospitals. Although such versatile medical services still exist in some industrialised countries, nowadays the occupational health services are mainly concerned with problems of prevention related to the work and the working environment (medical supervision of the health of workers and of conditions at workplaces). More rarely, the occupational health service may also engage in some aspect of general prevention (e.g. special check-ups for sportsmen, systematic electrocardiographic examinations for workers over 40 years of age, cancer check-ups for women workers, etc.).

In the developing countries, the situation is different. The demands for treatment may be so great that the establishment of health services related exclusively to prevention would be unthinkable, on the one hand because all the available medical resources must be pressed into service to meet the most urgent needs, and on the other hand because the case finding undertaken must be followed up by the provision of treatment to certain workers found to be suffering from a disease, deficiency or other abnormal condition. Since there is little possibility of administering it elsewhere than in the occupational health service, the only rational solution is for the latter to provide this care.

In mines, quarries, oilfields, plantations and building construction sites or public works in isolated areas, the occupational health service has an important task to play in helping to ensure the normal operation of the undertaking, and in some circumstances, in view of the lack of qualified workers and difficulty of replacing them, its role may be regarded as a vital one. In remote areas, these services will be made responsible for every aspect of health, including occupational health, medical treatment and care of workers and their families, and public health problems arising within the working community. It will have to concern itself not only with the treatment of occupational accident or disease victims, but also with child care, and with the provision of first aid and emergency treatment to the local population. It will have to provide obstetric care, and to see to the treatment and control of drinking water, the inspection of foodstuffs, the proper elimination of wastes and the provision of adequate facilities for workers. It will be responsible for vaccination campaigns, the medical examination of school children and apprentices, and the treatment and long-term supervision of patients with tuberculosis, trachoma, intestinal parasitosis, etc.

In referring to the examination of the different body systems, it was pointed out that many additional examinations found necessary will have to be performed and the administration of appropriate treatment seen to. How will this take place in practice?

In some countries, depending on the way in which medical services are organised, or in certain circumstances, the necessary additional examinations and treatment will be given by the occupational health service attached to the undertaking or common to several undertakings.

Fairly frequently, certain additional examinations are performed by the occupational health service, or on its initiative, in order to confirm a diagnosis, before an adolescent is sent to the physician who will give him the necessary treatment, and also to clarify the question of fitness for work. It should be observed that occupational health services are usually entirely free to perform or cause to have performed the necessary additional examinations required for diagnostic purposes. Sometimes the occupational health service confines itself to prevention, and the works doctor will advise the young worker to consult his own physician (and should also advise the parents that such a consultation is necessary). In such instances, it will be useful if the works doctor hands the young worker a note addressed to the physician who will be consulted, and if the two physicians exchange information and set up such co-operation as may be required by the case.

The young worker must be clearly aware of the possibilities afforded him in this respect, and therefore requires to have some idea of the organisation of the health services and how to go about using them. The works doctor should be in a position to supply information in this respect, and this implies that he himself should have a good knowledge of the existing possibilities in the region and of the professional and administrative organisation of medical services in his country, so that he can help the young person to derive the utmost benefit from the available resources.

Adaptation to Work and Job Satisfaction

Job satisfaction depends in a very large measure on the extent to which occupational life fulfils the hopes that have been placed in it. It is a function, therefore, not only of enrichment derived from the occupation and the occupational environment, but also of existing motivations, and of expectations fostered by the family, at school, and by the social and cultural *milieu*. It also depends on the extent to which the occupational environment fulfils the explicit or tacit fundamental needs of young people, to which reference has already been made. It is therefore obvious that factors making for satisfaction or dissatisfaction with the job vary widely, depending on the persons concerned, and that the relative importance seen as attaching to them is more variable still. For many young persons, remuneration is a main consideration, as representing objective evidence of participation in adult life. A young person who works to earn a living wants to earn enough to be able to live relatively independently, or to be able to help his family effectively, or to be able to save a reasonable sum towards marriage, for example. He wants to be treated equitably with respect to his adult workmates and to see the principle of "equal pay for work of equal value" translated into practice.

Many young workers appear to be very keen to participate effectively in useful and productive work involving close integration in a congenial working group where there is good understanding, and this factor contributes very substantially to their job satisfaction. Problems of authority are also very important for adolescents. It must be exercised firmly and justly, so as to guide them and enable them to progress; lack of authority, which leaves them to their own devices, is often disliked by them even more than its converse. An adolescent needs to have a social life, a life within a group, but the group must remain within human limits, and he will experience difficulty in fitting into large heterogeneous groups, in which personalities tend to become blurred and communication to become difficult, and the exercise of authority requires substantially higher pressure; small groups of five to ten persons appear to suit adolescents especially well, and it is worth noting that youth movements take this fact into account. Some adolescents set great store by the interest attaching to the work done, while others do not. Job satisfaction may be related to the use of certain materials or kinds of equipment; some young persons take pleasure in work with wood, metal or leather, while others like the sound and impression of power produced by their machines, while others still are keen to operate vehicles of all kinds. Some like to feel that they are creating something born of their work, while others are pleased when the machine they tend is operating efficiently, or when their work goes forward smoothly and output is high.

Reference must also be made to other factors which play a more or less significant role, such as hours of work and length of holidays, job stability, and opportunities for further training or promotion.

The conditions which make for job satisfaction may also lead to the contrary situation if they are not fulfilled. Where a specific ground for dissatisfaction exists, one not infrequently finds this dissatisfaction spreading to a variety of other fields. In particular, it may happen that working or environmental conditions which had previously been well tolerated begin to be regarded more and more as unsatisfactory, and complaints begin to pour in.

Job dissatisfaction may finally lead to lack of adaptation to work. The causes of this condition are so numerous that it is impossible to consider them systematically. Most of them are questions of psychology or character, but it must not be forgotten that they can also result from even quite minor physical or mental deficiencies or abnormalities, which should be detected sufficiently early, before the situation becomes irreversible. Although lack of adaptation to work is nearly always rooted in some kind of inner conflict, this does not mean that every such conflict necessarily leads to lack of adaptation; it will only do so when the conflict has been sufficiently prolonged and constitutes an unsurmountable obstacle, by reason either of its importance or of a particular weakness or deficiency or simply a lack of matureness evidenced by the adolescent concerned.

Lack of adaptation to work may find expression in different ways. Sometimes there may be a deliberate rejection of the work, occurring immediately or after some length of time; the person concerned loses all interest in his work and does only the minimum necessary to avoid sanctions and secure a reasonable remuneration. Sometimes there is instability in employment; the person concerned tries to overcome his general but vague sense of dissatisfaction through the quest for novelty, or he may change jobs in search of improved conditions, even though he may be unable to describe in specific terms just what he is looking for. Finally, there are situations which fall within the realm of professional failure. Frequently the person concerned goes through a transiticnal stage within which he tries to compensate and struggle against the growing inadaptation which he feels subconsciously, without being able to express it. In these circumstances, he will be able to maintain his position adequately only by exerting excessive pressure on himself, and tasks that he previously performed effortlessly demand increasing reflection and uninterrupted attention. The outcome may be increased fatigue, and - if vigilance falters - mistakes, mishaps and an increased tendency to be involved in occupational accidents.

Where signs of lack of adaptation are noted, the occupational health specialist should endeavour to pinpoint the situation of conflict which is responsible for it. He must try to obtain a clear picture, on the one hand, by studying the physique, psychology and character of the young person concerned, and, on the other hand, by going into his situation at the workplace and outside it, without neglecting any potential relevant factor. The importance of this task is seen clearly when one considers, on the one hand, the very serious risks for the personal, social and occupational future of an adolescent which may result from lack of adaptation to work, and, on the other hand, the fact that sometimes such a situation can be prevented from arising by quite simple means. Thus, the industrial physician may be led to give the young worker advice about how he should organise his work or behave at the workplace, or about the advantages of joining in certain spare-time cultural, sports or group activities. Sometimes further vocational guidance may be required, and sometimes the young person may have to be told to consult a specialist.

Young workers change their jobs especially frequently. This high degree of mobility is not altogether detrimental, since it enables young persons to gain an idea of the possibilities and conditions in various branches of industry before making a definite choice. If carried to excess, however, it can hamper the normal processes of qualification and advancement, and in particular, it can sometimes pose substantial problems in undertakings. This mobility reflects to some extent the wish of young persons to learn about life, to broaden their horizons and acquire real independence. But very frequent changes of jobs can also result from continued dissatisfaction or a lack of adaptation to the work or even to the social environment; in such cases, the causes must be determined before the necessary remedial action can be taken.

Role of the Occupational Health Service

Before winding up this survey of the role and responsibilities of industrial physicians with respect to young workers, it is appropriate to consider the means by which the important and delicate task of supervising the health of adolescents at the level of the undertaking can be carried out.

ILO Recommendation No. 112 concerning Occupational Health Services in Places of Employment defines the role of these services as that of protecting the workers against any health hazard which may arise out of their work or the conditions in which it is carried on, and contributing towards the adaptation of the work to the workers and to the establishment and maintenance of the highest possible degree of physical and mental well-being in them.

At the administrative level, these services may be organised in various ways, to which the Recommendation refers. Under the heading of functions, the Recommendation also reviews in detail the many activities of occupational health services, ranging from surveillance of hygiene at workplaces, job analysis and participation in the prevention of occupational accidents and diseases, to the field of preventive medicine in all its aspects, such as medical examinations, surveillance of adaptation of jobs to workers, medical advice to workers, emergency and ambulatory treatment, and education of personnel in health and hygiene, etc.

As a result of variations in national practice, the manner in which these services are organised and the scope of some of their functions may vary to some extent from one country to another. But the most critical factors in this respect would appear to be the training and attitude of the responsible physician, which must be such as to enable him to exercise his activity in such a way that it will contribute as constructively as possible to the harmonious development of adolescents. And indeed, throughout this publication, stress has been consistently laid on the dominant role played by the industrial physician in this respect.

ILO Recommendation No. 112 provides that the physician in charge of an occupational health service should have received, as far as possible, special training in occupational health, or at least should be familiar with industrial hygiene, special emergency treatment and occupational pathology, as well as with the laws and regulations governing the various duties of the service, and that he should be given the opportunity to improve his knowledge in these fields. In order to be able to help young workers effectively, the industrial physician must naturally be acquainted with their trades and occupations, the capacities required and risks involved, and must be familiar with and regularly inspect their workplaces. But he will also be called upon to deal with psychological problems and questions of human relations; he will have to take an interest in problems relating to education, the working environment, vocational guidance, prevocational and vocational training, the introduction and integration of adolescents in the undertaking and at the workplace, as well as the nutrition, physical education and housing of young workers. Experience has shown that very frequently a knowledge of all these problems, including psychological and psycho-social problems, is indispensable to the industrial physician in finding the answers to certain questions that concern him directly. Disappointed hopes, poor social adaptation and a mistaken picture of occupational life may play an important part in the occurrence of chronic fatigue, otherwise inexplicable loss of weight and repeated accidents. Sometimes latent anxiety stemming from unexpressed fears due to circumstances or disputes arising at the workplace may be responsible for digestive disorders, excessive cardiac and circulatory system reactions or high blood pressure. If too narrow a view is taken of these problems, the physician may fail to attribute certain disorders found at a medical examination to their true causes.

Although the industrial physician must have a due knowledge and understanding of all the problems that may arise in connection with the medical supervision of young workers and keep them constantly in mind, he will not necessarily be responsible for dealing directly with each and every aspect of them. The division of work in an occupational health service depends on the physician and the on the way in which he wishes to organise his service, and may vary, depending on circumstances and on the availability, training and competence of staff.

Many duties may well be performed by a medical assistant or nurse. These include, typically, the initial reception of young workers, the provision of information about the role of the occupational health service and the regular inspection of washrooms, cloakrooms and refectories and of working clothes, as well as (after appropriate treatment) the periodic briefing of every young worker about the occupational hazards that he may encounter at his workplace, and any health measures required. Indeed, such a person may be in a better position - as is frequently the case in practice - to establish a favourable atmosphere of mutual trust and understanding.

What matters is not that the physician should do everything, but that all the responsibilities of the occupational health service should be duly discharged by persons who are aware of the importance of their role and take it to heart, that the physician should be kept informed of what is being or is to be done, and that he should assume personal responsibility for the efficient performance of all the different kinds of work that a well-organised occupational health service must undertake.

ILO Recommendation No. 112 provides that the nursing staff attached to occupational health services should possess qualifications according to the standards prescribed by the competent body, and that the first-aid personnel should consist exclusively of suitably qualified persons.

In order to ensure that the physician can work efficiently and that his time is not wasted on a series of tasks for which his high qualifications are not necessary, adequate staff must be made available to him. This is a mandatory condition not only in the industrialised countries, but perhaps in particular in the developing countries, where the shortage of specialised personnel is sometimes very acute; in these countries, one of the essential tasks of the physician may then consist in the selection and training of his staff, which may at the outset lack most of the necessary qualifications.

Another important role of the physician is that of establishing and maintaining close relations with the other services and bodies in the undertaking concerned with occupational safety and health and the welfare of workers, including in particular the social welfare department, the safety department, the personnel department, trades union committees in the undertaking, and the safety and health committee or any other committee or persons concerned with safety and health questions in the undertaking; he must also encourage the workers and their organisations to give the occupational health service the benefit of their full co-operation in the pursuit of this aim.

The physician must also maintain contact with services and bodies outside the undertaking dealing with problems of occupational safety and health and with the retraining, rehabilitation, resettlement and welfare of workers. Particularly close contact should be maintained with the medical branch of the labour inspectorate. On some occasions, the works doctor may be asked to co-operate in certain general studies of various subjects, while in others, he may seek the backing of the medical inspector of labour in pressing for certain improvements or changes, and in both cases, some of the questions touched on may be of relevance for young workers.

CHAPTER IV

HISTORICAL SURVEY OF NATIONAL AND INTERNATIONAL LEGISLATIVE ACTION TO PROTECT THE HEALTH AND WELL-BEING OF YOUNG WORKERS

National Laws and Regulations

Prior to the period of profound technological, economic and social transformations which came to be known as the "Industrial Revolution" and spanned the end of the eighteenth and the greater part of the nineteenth century, hardly any legislation had been promulgated regulating work by children and young persons. Although the conditions of work in the different crafts and in occupations pursued at home were far from favourable, at least these young persons frequently worked side by side with their parents. On the land, the continued existence of conditions which were virtually feudal in nature led to the exploitation of large numbers of agricultural workers and of their families, even very young children being sent to work. The work of apprentices, however, were subject to certain rules and regulations enforced by the responsible guilds and governing apprenticeship to the trades concerned.

The introduction of machinery revolutionised not only production processes but also working conditions. Work which had hitherto been performed at home or in small workshops began to be centralised in factories, which in the early stages of industrialisation were mostly located along the waterways, and were subsequently set up in the towns when steam superseded hydraulic energy as a source of motive power.

The exploitation of children began when it was found that they could be employed to operate machines which were performing operations which had previously been done laboriously by hand. The conditions of employment in the newly established factories were not subject to any laws or regulations whatsoever; moreover, the employers, being anxious to recoup the capital laid out in the purchase of the new machines as quickly as possible, tried by every means in their power to squeeze the maximum possible amount of labour and output from their workers. Orphans and other penniless children maintained at the expense of the parish were often sent to work in the factories, where they found themselves in a sorry plight, as the children of poor families did later on, when the lot of the former had been improved by legislation.

As a result of the increasing need fcr coal as fuel for steam plant, mining developed rapidly, and many children were also set to work in the mines where they laboured hard in conditions which were not only dangerous but also very damaging for their health and development.

It was in the United Kingdom, where the industrial revolution took place in the textile industry earlier than in other countries, that the first laws and regulations regulating work by children and young persons were adopted and implemented. Legislation with similar aims was introduced in France somewhat later. The French and British provisions inspired subsequent action in this field by other European countries.

This early legislation appears to have been motivated primarily by the wish to ensure that the young workers should receive some instruction, which led the legislators to fix a minimum age and period of attendance at school as conditions for admission to employment. Although their scope was limited and their application left much to be desired, these texts nevertheless laid down a principle which, once it became established, resulted in the legislation being successively widened to cover almost every branch and kind of employment in the broadest sense.

The laws and practices in different countries, however, tended to develop along different lines, and from the beginning of the twentieth century onwards, efforts were made to guarantee equitable working conditions by the adoption of international instruments.

The International Labour Organisation was established in 1919, and the First Session of its Conference was held in the same year. The international instruments adopted at that session included certain Conventions and Recommendations relating to the work of children and young persons. Subsequently, the International Labour Conference adopted many other Conventions and Recommendations, designed to serve as models or guidelines for national laws and regulations. The standards laid down by the international Conventions and the guidelines established by the Recommendations have very substantially influenced the development of national legislation relating to the employment of children and adolescents.

Many of these Conventions have been ratified by a large number of ILO member States and their provisions have been taken up in the respective national laws and regulations.

In addition, and independently of the aspect of ratification and implementation, the standards thus established have played an important role, even in countries which have not ratified these Conventions, by influencing the broad trends of public opinion and the development of national legislation in these countries.

United Kingdom. The occurrence of occupational cancer in chimney sweeps was reported in 1775, and the employment of young children in such work was formally banned by the Chimney Sweep Act of 1788 (although the employment of boys in this trade continued for a number of decades thereafter).

Other accounts testifying to the serious health risks incurred by child and adolescent workers began to appear from about 1770 onwards; yet it was not till 1802, after public opinion had been aroused by a major epidemic which had broken out among pauper apprentices in the north of England, that Parliament adopted an "Act for the Preservation of the Health and Morals of Apprentices and Others Employed in Cotton and Other Mills". This Act, which applied only to the cotton textile industry and to parish apprentices, provided that the hours of work of apprentices were not to exceed 12 a day (not including mealtimes), and that apprentices were not to be employed on night work (i.e. between 9 p.m. and 6 a.m.), that rooms in factories were to be lime-washed, and that the apprentices were to be given adequate clothing, religious instruction and general education.

The immediate and totally unexpected result of this legislation was the employment of "free" child labour, that is to say, of the children of impoverished parents (in contrast to pauper children made available by the parishes), whose employment was not subject to any rules or regulations. Thus the situation quickly became as bad as before.

In 1819, a second Act was passed which applied to all children working in the cotton textile industry. This Act, which fixed 9 years as the minimum age for admission to employment, and 12 hours as the maximum length of the working day, remained practically a dead letter, as it contained no provision for its implementation. Nevertheless, it at least had the merit of laying down the principle that child labour should be regulated by law. In other Acts of 1825 and 1831, further attempts were made, with little success, to improve the situation, and it was only in 1833 that the first really effective Factory Act was passed, which prohibited the employment of children less than 9 years old in certain textile factories, and limited the daily hours of work to 9 for children less than 13 years old and 12 for young persons under 18 years of age, who were also not allowed to do night work, or to work more than 69 hours a week. This Act also contained provisions designed to ensure that young workers received an education, by requiring all child workers to have a certificate, made out by a schoolteacher, certifying that the child was receiving at least the minimum education required for admission to employment. It also provided for the appcintment of four regional factory inspectors to report on its implementation.

In 1844, another Act was passed extending the regulation of child labour to a larger number of textile factories. Although reducing the minimum age of admission to employment to 8 years, this new Act also cut the length of the working day to 6 1/2 hours for children less than 13 years old, and further provided that this work must be done either before or after the midday meal, thus in effect instituting a half-time system under which children worked for half of the day and attended school during the other half. Other provisions prohibited employment of children in work with certain dangerous machines, required the age of the young wcrker to be attested to by a certificate delivered by a "certifying surgeon", and further required every child to have a certificate of instruction made out in a set form, this being necessary to stop the abusive practice of certifying instruction that had not been given, which had sprung up under the previous legislation. This Act represented a substantial advance compared to the 1833 Factory Act, by tightening up the provisions relating to age of admission to employment and the education of young workers.

In 1845, Parliament enacted the Print Works Act, laying down various requirements concerning the age of admission to employment, working hours, night work, medical certificates and attendance at school, as well as provisions to ensure its implementation.

In 1860, textile dyeing and bleaching shops were made subject to the provisions of the Factory Act.

At this time, regulaticns governing the employment cf children and young persons were still limited to the textile industry, although that industry was by no means the only one to make use of child and adolescent labour. Many children were also working in open-air brickworks, potteries, glassworks, metal-grinding works and even in the production by dipping of matches tipped with white phosphorus. Indeed, the restrictions imposed by the Factory Act drove children to seek employment in branches of industry not subject to these controls. As a result, an Act was passed in 1864 extending the provisions of the Factory Act to children employed in six other branches of industry as well as the textile industry, and in 1867, with the adoption of the Factories and Workshops Act, the employment of children was controlled in a number of previously unregulated industries, including foundries and certain other metalworks, and a variety of other dangerous or unhealthy occupations; a further Act subjected factories and shops employing less than fifty workers to the provisions of the Factory Act. Thus many further categories of child and adolescent workers were protected by the provisions of these two Acts.

From that time onwards, British industrial legislation progressively raised the minimum age of admission to employment, which finally reached 14, while at the same time further restricting night work by young persons, and strengthening the provisions concerning education of young workers. The system of half-time working, with children working in a factory for half the day and going to school during the other half, was abandoned in 1920.

The conditions of employment of children in mines were first regulated in 1842 with the passage of the Mines Act, which prohibited the employment of boys under 10 and of girls of any age in underground work in mines of any kind. In 1872, the minimum age for admission to employment was raised to 12 years by Acts relating respectively to coal mines and metalliferous mines (this distinction being observed in subsequent legislation). In 1911, the minimum age for employment in underground work was raised to 14 years.

France. Until 1840, a 14-hour working day was enforced throughout industry both for adults and for children. No minimum age for employment existed until 3 January 1813, when Napoleon issued a decree setting the minimum age at 10 years. The first piece of legislation governing work in industry was the "Act of 1841 concerning the work of children employed in factories", which applied to factories or workshops using mechanical engines or continuous firing, as well as to all factories employing more than 20 workers in a workshop. It set the minimum age for admission to employment at 8 years, and fixed a working day of 8 hours for children between 8 and 12 years old. It prohibited the employment of young persons between 12 and 16 years of age for more than 12 hours daily, or at night (i.e. between 9 p.m. and 5 a.m.), except in emergencies, and placed a total ban on night work of children under 13 years of age. The age of the children had to be confirmed by a certificate delivered by an official of the registry office; children under 16 years of age could not be employed on Sundays or holidays, and all children admitted to employment had to go to school until the age of 12. The Act also provided that factory managers should keep records of the children employed, each of whom should be provided with a work book. Implementation of the Act was to be supervised and enforced by inspectors who could call on each establishment and ask to see the employment records and work books; employers failing to comply could be penalised by fines. Public opinion was indifferent to the new Act, which aroused hostile reactions on the part of the workers, whom it was supposed to protect, but who regarded it chiefly as a measure designed to decrease or abolish the earnings of their children. Its provisions were flouted everywhere; children under the legal age continued to be employed in factories or workshops, the ban on night work remained inoperative, and the system of inspections did not produce the anticipated result.

Effective protection was secured only with the passage of the Act of 1874, which fixed the minimum age of admission to regular employment in a fairly wide variety of industrial establishments at 12 years, while allowing children between 10 and 12 years old to be employed half-time (6 hours daily) in some branches of industry as an exceptional measure, and setting a maximum working day of 12 hours for adolescents between 12 and 16 years old. The Act prohibited night work of boys under 18 years of age and girls under 21 years of age, except in factories using continuous firing, guaranteed a day of rest for children and young persons on Sundays, and prohibited underground work of boys less than 12 years old and minor girls of all ages. Adolescents under 16 years old could not be employed on certain types of work regarded as dangerous or unhealthy. Twelve-year old child workers were required to produce a medical certificate confirming their fitness for employment in the job foreseen (this new provision was modelled on British, Italian and Scandinavian legislation). Last but not least, penalties were imposed and effective arrangements made for inspection.

But the 1874 Act was still insufficient, inasmuch as it did not apply to workshops attached to charitable institutions or vocational schools or ministries, etc., and young male workers over 16 years of age were not protected. Another Act was accordingly passed on 2 November 1892 to regulate work by children and women. This Act extended the provisions on child labour to workshops not covered by the Act of 1874, fixed a maximum working day of 10 hours for children under 16 years of age and 11 hours daily or 60 hours weekly for adolescents from 16 to 18 years of age, a working day of 12 hours being fixed for adult men. In addition, the Act prohibited underground work in mines by boys under 13 years of age and girls of all ages. The admission of adolescents from 13 to 16 years old to work in mines was made subject to various conditions.

In 1900, the drawbacks resulting from the setting of working days of different lengths for various categories of young workers led to the passing of an Act fixing a working day of 11 hours for all workers, with provision for reduction to 10 hours in successive stages, in 4 years' time. These Acts were consolidated in 1912 and incorporated in the Labour Code (Book II); the minimum age of admission to employment was set at 13 years, or 12 years for children with a certificate of elementary education.

Other European countries. All European countries have enacted legislation regulating the employment of young persons (in many cases it has been in existence for a long time), and covering the same broad aspects, such as age of admission to employment, night work, hours of work, weekly rest and annual holidays, and prohibition of underground work and of certain kinds of work involving specific hazards.

In Sweden, for example, the employment of children on night shifts in factories has been prohibited since 1852. In 1881, legislation was adopted for the protection of labour which, amongst other things, prohibited the employment of children under 12 years of age in factories, and instituted a shorter working day, as well as requiring employers to be in possession of certificates indicating the age and state of health of the worker and giving details of school attendance. After about 1880, Swedish legislation in this field developed very largely along the same lines as in the United Kingdom.

In Belgium, the laws and regulations have been modelled substantially on the French legislation. Although a bill was drafted as early as in 1848 to limit the working day of children and adolescents from 10 to 18 years of age to 6 1/2 hours daily and to restrict work of women and children in underground mine workings, and despite several moves by managers of coal mines and other industrial undertakings, it was only in 1884 that a royal order was finally promulgated fixing the minimum age for admission to employment at 12 years for boys and 14 years for girls. In 1889, an Act on the employment of women and children was adopted establishing a maximum working day of 12 hours for boys between 12 and 16 years of age and girls up to 21 years of age, prohibiting night work and providing for a day of weekly rest for these categories of workers, and prohibiting the employment underground of girls under 21 years of age. In 1911, the prohibition of underground work was extended to all women and to boys under 12 years of age, the minimum age limit for such work being raised to 14 two years later.

United States. In the United States, the industrial revolution took place later than in the United Kingdom, but it had somewhat similar consequences, except that child labour was not exploited on such a wide scale. In that country, legislation in this field has always been a matter for the states rather than for the federal Government. The first laws enacted for the purpose of regulating the employment of children and young persons made inadequate provision for implementation. As in the United Kingdom, the first evil to be tackled was the lack of education of young workers, and some states enacted laws to remedy this state of affairs. The first state to pass legislation on child labour was Connecticut, where a law enacted in 1813 required factory owners to have the children employed by them taught reading, writing and arithmetic.

In 1836, the State of Massachusetts required working children under 15 years of age to attend school for 3 months a year. Similar laws relating to schooling were passed in other states. Finally, in order to be admitted to employment, children were required to have completed a year of schooling at a given standard.

In 1842, the State of Massachusetts limited the hours of work for children under 12 years old to 10 a day in some branches of industry; in the same year, the State of Connecticut passed a similar law, and by 1860, 6 states had adopted analagous legislation.

Pennsylvania was the first state to bar the employment of very young children, by passing a law in 1848 setting a minimum age of 12 years for employment in factories. Gradually, the other states adopted similar regulations concerning children working in factories. In 1930, two states - Montana and Ohio - imposed a minimum age of 16 years for admission to employment; by 1938, 8 other states had also adopted this standard.

Provisions to protect the health of children were enacted in legislation empowering labour inspectors to prohibit the employment of children whom they considered to be physically unfit to perform the work allotted to them. Furthermore, every child in doubtful health was required to undergo a medical examination before a work permit could be delivered for him. Further legislation required all children to be examined by a government physician before taking up employment, which they could do only if they were passed as physically fit for the work.

The legislation on child labour also tended to protect young workers from employment in dangerous and unhealthy occupations. Initially, this aspect was dealt with in the penal legislation, but subsequently it was covered by the labour laws, and labour inspectors were made responsible for enforcing the relevant provisions. Towards 1900, schedules of machinery, the use of which was prohibited in undertakings employing young workers, and of unhealthy occupations began to be introduced in the laws on child labour. The federal Government made several attempts to co-ordinate the laws and regulations enacted by the various states to regulate the employment of children and adolescents. However, federal legislation on child labour, although initiated in 1916, did not make significant progress until 1938, when Congress passed the Fair Labour Standards Act, which defines oppressive child labour as the employment of minors under 14 years of age. Children between 14 and 16 years old may be employed outside school hours in a limited number of occupations, when such employment does not interfere with their schooling, health or well-being.

USSR. At the outset of its existence, in 1919, the Soviet Union passed a decree doing away with child labour.

Soviet legislation forbids the admission to employment of adolescents under 16 years of age. Young persons of 15 full years of age may be employed only in certain circumstances, with the specific agreement of the local trade union committee or the trade union committee of the undertaking.

In October 1922, a schedule of dangerous or unhealthy occupations, from which young persons under 18 years old are banned, was adopted; this schedule has undergone subsequent revision. Schedules indicating the minimum age of admission to employment and any restrictions based on sex, and schedules of affections entailing unfitness for work, are also laid down in the legislation in respect of a whole range of occupations in all sectors of industry and trade.

Legislation requiring the compulsory medical examination of adolescents for employment was adopted in 1922; subsequent periodic examinations are also mandatory. The purpose of this medical supervision is to detect at the outset any signs indicating that the work is having an adverse effect on the worker's health, so that the necessary steps can be taken to select more appropriate work for him (change of workstation, redirection with fresh vocational training, etc.).

Soviet legislation restricts the hours of work of young workers, so as to enable them to enjoy adequate rest and to continue their studies or receive additional vocational training; towards this end, a reduction of 2 hours is made in the working day, or of 2 days in the working week. In addition, young workers are entitled to additional leave for the purpose of taking examinations or tests or preparing for diplomas.

Young persons of 15 and 16 years of age undergoing individual or group apprenticeship work a 4-hour day. A decision adopted by the Council of Ministers of the USSR in August 1955 prohibits young adolescents from doing overtime or night work, as well as from working on days of rest or on holidays.

Young workers under 18 years of age are entitled to an annual holiday of one month, provision being made for them to take this holiday in summer, as being the best time of year for rest and relaxation.

Other countries. In general, the laws and regulations relating to the protection of children and young persons do not go back much beyond the beginning of the twentieth century. Sometimes, indeed, they are more recent still, in the case of States that have not long been independent, although conditions in these territories were sometimes regulated previously by provisions very similar to those applying in the metropolitan countries, such as the Labour Code applying to territories coming under the jurisdiction of the Ministry for French Territories Overseas.

Examples of essential features of certain national laws and regulations follow.

In Senegal, the population figures as of 1 January 1969 show that at that time 60 per cent of the 3,780,000 inhabitants were less than 24 years old; problems of youth are therefore a main preoccupation of the Government. Act No. 61-34 of 15 June 1961 establishes the nature and form of the contract of apprenticeship; it provides that children may not be employed in any undertaking, even as apprentices, before the age of 14 years, except on the strength of a waiver granted by order of the Minister of Labour and Social Security, issued after consulting the National Advisory Council for Labour and Social Security, and bearing in mind local conditions and the nature of the work involved. An order issued by the Minister of Labour and Social Security lays down the kinds of work and undertakings in which young persons may not be employed, and the age limits below which this prohibition applies. Special provisions also cover pre-employment and periodic medical examinations. Reference should also be made to the existence of institutions and movements designed to prepare young Senegalese who are not attending school or undergoing apprenticeship for a working life, namely, in particular, the "educational work camps" designed to secure the participation of young persons in the development of "pioneer zones", and the "national civic service", which is organised along military lines and is designed to "provide a number of young persons between 16 and 21 years old with a solid civic and moral training, and to introduce them to modern methods of agriculture and fishing, and to various specialised crafts (woodwork, metalwork, masonry)".

In Tunisia, the Decree of 31 December 1907 setting up the Labour Office provided that the latter would be responsible in particular for ensuring adequate occupational safety and health conditions for workers, especially women and children. With two decrees issued in 1910, a first step was taken towards the restriction of child labour and the regulation of conditions of work (light work and underground work prohibited for women and girls and restricted for boys under 16 years of age, regulation of apprenticeship). Between 1939 and 1965, more decrees were issued, and in 1966, Act No. 66-27 instituted the Tunisian Labour Code, which, with certain additional decrees, provides the legal foundation for the medical supervision and health protection of young workers. In this country, the minimum age for employment is set at 15 years in industry and 13 years in agriculture and other light work; special provisions apply to vocational training. The working day is limited to a maximum of 2 hours for children under 14 years of age, 4 1/2 hours for children of 14-15 years of age, and 6 hours for young persons between 15 and 18 years of age, for employment on work which is neither industrial nor agricultural. Night work, underground work and various other kinds of work, including in particular very arduous or dangerous work in agriculture, are prohibited. Paid annual holidays are doubled (except in agriculture) up to the age of 18 years, and increased by half between 18 and 21 years. Provision is made for the medical supervision of young workers, which may be required by the Labour Inspectorate.

In Japan, the employment of women and children became substantial only after the country had been industrialised in the late nineteenth and early twentieth century. In 1911, in spite of the opposition of the employers, an Act, the purpose of which was mainly to protect women and children, was adopted. This Act, which entered into force in 1916, prohibited the employment of children under 15 years of age (16 years since 1923) for more than 12 hours daily (11 hours in 1923); it also prohibited night work between 10 p.m. and 4 a.m. (5 a.m. in 1923) and some kinds of dangerous work; young persons under 15 years of age only (16 years in 1923) were to enjoy 2 days of rest, and a work break of at least half-an-hour during working hours, when the latter exceeded 6 hours, and at least 1 hour, when they exceeded 10 hours daily; provision was also made for compensation of occupational accidents or diseases. The Act applied only to undertakings employing more than 15 workers (10 workers in 1923). In 1923, the minimum age for admission to employment in maritime work was set at 14 years. Analogous provisions were contained in the Mines Act of 1905 (amended in 1926). Until 1946, relatively little progress was made, and the Japanese legislation remained very inferior to the international standards. The year 1947 witnessed the adoption, following strong pressure by the trade unions, of an important item of legislation, the Labour Standards Law, which completely changed the situation in Japan and laid down very high standards for the protection of women and children: the minimum age for admission to employment is fixed at 15 years (with rare exceptions), with provisions for special protection until 18 years; a ban is placed on the employment of young persons under 18 years of age in night work (10 p.m. to 5 a.m.) and different kinds of dangerous or unhealthy work; the working day is fixed at 8 hours (possibly 10 hours for young persons between 15 and 18 years of age, if only 4 hours are worked on another working day of the week and the working week does not exceed 48 hours); provision is also made for 48 hours of weekly rest and a daily work break of 45 minutes; overtime is prohibited. Section 52 of this Law provides that a medical examination of fitness and a dental examination (made by a dentist) must be performed at the time of employment and thereafter periodically. The Labour Inspectorate is responsible for supervising the implementation of this Law, pursuant to which various Orders have also been issued, including in particular an Order establishing a schedule of occupations in which young persons may not be employed.

In Iran, the 1959 Labour Code prohibits the employment of children under 12 years of age, even for purposes of apprenticeship. Night work (10 p.m. to 6 a.m.) by women and young persons under 18 years of age is prohibited (with a few exceptions, such as hospital nurses). The employment of young persons under 18 years of age in arduous or dangerous occupations is also prohibited.

International Standards

Action by the International Labour Organisation on behalf of young persons has in the first place taken the shape of the drafting and adoption of international standards laying down the principles which should be reflected in national action to protect children and adolescents, by safeguarding their health and physical and mental development and ensuring that they receive education and training and are given employment suited to their age.

While the entire body of ILO standards applies no less to young workers than to adult workers, some standards also contain special provisions relating to the former, while others are specifically aimed at the protection of young persons. These include in particular the standards respecting, the minimum age for admission to employment in certain occupations involving risks to the health, night work, and medical examinations for fitness for employment. Reference should also be made to the Resolution concerning the Protection of Children and Young Workers, adopted in Paris in 1945, which is a real "young persons' charter"; this Resolution, which considers the problem of young persons in all its aspects and details the measures that should be taken on behalf of young workers, marks the starting-point of a new and more dynamic approach to these problems.

In addition to its standard-setting activities, the ILO has acted as a clearing-house for information from all parts of the world on the problems of young persons, which it collects and then makes available systematically in the shape of studies, surveys, reports and other publications. In this way, a vast store of data is made available to member States. Many governments have requested and obtained the assistance of the ILO in helping them to draft their legislation concerning young persons and various aspects of the vocational training of adolescents and the protection of young workers. In the last few years, within the framework of its technical assistance programmes, the ILO has been increasingly active in assisting governments to solve the practical problems facing them in this field.

The problems of young workers have also been considered by meetings of experts and of various ILO bodies, such as the Panel of Consultants on Young Workers' Problems, the Industrial Committees and the Permanent Agricultural Committee, and have likewise been discussed at the Regional Conferences. At these various meetings, a large amount of information has been collected about experience of particular interest obtained in various countries, and an effort has been made to establish guidelines and define appropriate methods of dealing with these problems.

At the international level, the ILO co-operates closely with the other organisations concerned with the problems of young persons. Such co-operation, for example, is pursued by the ILO and UNESCO with respect to education and vocational guidance and training, while the ILO and WHO co-operate in matters concerned with the help of adolescents, maternity protection and health protection for women and children; again, the ILO co-operates with FAO on questions affecting young persons in rural areas, and with the United Nations as regards the social problems, training and protection of young persons.

The following is a brief survey of the most important international instruments relating to the protection of young workers (where instruments have been revised, reference is made to the most recent text).

Minimum Age for Admission to Employment

Agriculture. ILO Convention No. 10 of 1921 concerning the Age for Admission of Children to Employment in Agriculture provides that children under the age of 14 years may not be employed or work in any public or private agricultural undertaking, or in any branch thereof, save outside the hours fixed for school attendance. If they are employed outside the hours of school attendance, the employment shall not be such as to prejudice their attendance at school. Nevertheless, for purposes of practical vocational instruction, the periods and hours of school attendance may be so arranged as to permit the employment of children on light agricultural work, and in particular on light work connected with the harvest, provided that such employment shall not reduce the total annual period of school attendance to less than eight months.

Employment at sea. ILO Convention No. 58 (Revised), 1936, Fixing the Minimum Age for the Admission of Children to Employment at Sea provides that children under the age of 15 years shall not be employed or work on vessels, other than vessels on which only members of the same family are employed. However, national laws or regulations may provide for the issue in respect of children of not less than 14 years of age of certificates permitting them to be employed in cases in which an educational or other appropriate authority is satisfied, after having due regard to the health and physical condition of the child, that such employment will be beneficial to him.

ILO Convention No. 112 of 1959 concerning the Minimum Age for Admission to Employment as Fishermen provides that children under the age of 15 years shall not be employed or work on fishing vessels. However, such children may occasionally take part in the activities on board fishing vessels during the school holidays, provided that such activities are not harmful to their health or normal development, are not such as to prejudice their attendance at school, and are not intended for commercial profit. Furthermore, national laws or regulations may provide for the issue in respect of children of not less than 14 years of age of certificates permitting them to be employed on board fishing vessels, in the same conditions as are laid down in Convention No. 58.

Industrial employment. ILO Convention No. 59 (Revised) of 1937 fixing the Minimum Age for Admission of Children to Industrial Employment provides that children under the age of 15 years shall not be employed or work in any public or private industrial undertaking, or in any branch thereof. However, national laws or regulations may permit such children to be employed in undertakings in which only members of the employer's family are employed, provided that such employment is not dangerous to the life, health or morals of the persons employed therein. ILO Recommendation No.

52 of 1927 concerning the same subject recommends that member States should make every effort to apply their legislation relating to the minimum age of admission to all industrial undertakings, including family undertakings, by suppressing this exception. Special provisions fixing lower minimum ages for admission to employment are applicable in Japan and China.

Non-industrial employment. ILO Convention No. 6o (Revised) of 1937 concerning the Age for Admission of Children to Non-Industrial Employment provides that children under 15 years of age, or children over 15 years who are still required by national laws or regulations to attend primary school, shall not be employed in any employment to which the Convention applies, with the following exceptions: children over 13 years of age may, outside the hours fixed for school attendance, be employed on light work which is not harmful to their health or normal development and is not such as to prejudice their attendance at school. National laws or regulations shall specify what forms of employment may be considered to be light work, and prescribe the number of hours per day during which it may be performed, except that no child under 14 years of age may be employed on such work for more than 2 hours per day. In countries where no provision exists relating to compulsory school attendance, the time spent on night work shall not exceed 4 1/2 hours per day. Special provisions fixing lower minimum ages are applicable in India.

ILO Recommendation No. 41 of 1932 concerning the same subject emphasises the desirability of restricting the employment of children to as great an extent as possible as long as they are required to attend school, in order that their physical, intellectual and moral development may be safeguarded. In addition, for the admission of children to employment in light work, the competent authorities should require the consent of parents or guardians, a medical certificate of physical fitness for the employment, and, where necessary, previous consultation with the school authorities.

Hours of Work

Hours of work of young persons in full-time employment. The various ILO Conventions regulating hours of work make no distinction as to age, and apply with equal force to young workers and to adult workers. There are no international standards providing for any special reductions in hours of work for young workers in full-time employment.

The first international labour Convention adopted by the International Labour Conference at its First Session in 1919 limited the hours of work in industrial undertakings to 8 in the day and 48 in the week. In 1930, a further Convention was adopted applying the same limits to hours of work in commerce and offices.

The next stage was to consist in a reduction in the hours worked per week. Indeed, the 40-hour week had become one of the main objectives of the trade unions. In 1935, the Conference adopted a Convention upholding the principle of a 40-hour working week.

The question of hours of work of seamen is dealt with in several Conventions concerning wages, hours of work on board ship and manning. Two Recommendations adopted in 1920 called for the introduction of an 8-hour working day or a 48-hour working week in the fishing industry and in inland navigation. Convention No. 46 (Revised) of 1935 lays down special provisions limiting the hours of work in coal mines. Convention No. 67 of 1939, for its part, contains a series of special provisions regulating hours of work and rest periods in road transport. Several Conventions adopted between 1934 and 1937 provide for reductions of hours of work in various sectors of industry, in particular in glass and bottle works, in the textile industry and in public works.

The Resolution adopted by the International Labour Conference in Paris in 1945 supported the principle of a special reduction in the daily and weekly hours of work of young workers, which should be strictly regulated to take into account the needs of the different age groups; it recommends that, to the extent possible, the working week of young persons and children not attending school should be reduced and should not exceed 40 hours.

Hours of work of young persons undergoing vocational training. The Conventions on minimum age for admission to employment are not applicable to young persons undergoing vocational training, whose activities during that period must be approved and supervised by the public authorities. The same applies to the international standards regulating hours of work. In a very general way, it seems that the laws or regulations on the protection of young workers are not applied during the period of vocational training, and this is sometimes a source of problems for which an effective solution should be found as quickly as possible.

ILO Recommendation No. 57 of 1939 concerning Vocational Training provides that the time spent in attending supplementary courses by apprentices and other young workers who are under an obligation to attend such courses should be included in normal working hours.

The Resolution adopted by the International Labour Conference in Paris in 1945 calls for the making of appropriate arrangements during working hours so as to permit attendance at continuation courses of general or technical education, by setting a legal maximum on the aggregate hours of school and work, these hours being preferably remunerated as working time.

ILO Recommendation No. 117 of 1962 concerning Vocational Training provides, further, that measures should be taken to ensure that the conditions of work of persons, particularly young persons, who are receiving training, whether in an undertaking or a training institution, are satisfactory, and in particular that the work done by them is suitably restricted so that it is essentially of an educational character.

Overtime. As the systematic working of overtime could make the regulation of hours of work inoperative, this question is usually also regulated by suitable legislation. For the work force as a whole, restrictions on overtime working place a ceiling on daily, weekly, monthly or yearly hours of work, alone or in combination, calculated by applying more or less complex formulae. The majority of countries have adopted laws and regulations prohibiting overtime working by young workers, except in special cases.

ILO Recommendation No. 116 of 1962 concerning the Reduction of Hours of Work stresses that in arranging overtime due consideration should be given, amongst other things, to the circumstances of young persons under 18 years of age.

Night Work

Agriculture. ILO Recommendation No. 14 of 1931 concerning Night Work of Children and Young Persons in Agriculture provides that each member State be asked to take steps to regulate the employment of children under the age of 14 years in agricultural undertakings during the night, in such a way as to ensure to them a period of rest compatible with their physical necessities and consisting of not less than 10 consecutive hours. For young persons between the ages of 14 and 18 years, the period of rest must consist of not less than 9 consecutive hours.

Employment at sea. ILO Convention No. 109 (Revised) of 1958 concerning Wages, Hours of Work on Board Ship and Manning provides that no person under the age of 16 years shall work at night; the term "night" means a period of at least 9 consecutive hours between the times before and after midnight to be prescribed by national laws or regulations, or collective agreements.

Industrial employment. ILO Convention No. 90 (Revised) of 1948 concerning the Night Work of Young Persons Employed in Industry provides that young persons under 18 years of age shall not be employed or work during the night in any public or private industrial undertaking or in any branch thereof. However, for purposes of apprenticeship or vocational training, the competent authority may authorise the employment in night work of young persons between 16 and 18 years of age; in such cases, the young persons concerned shall be granted a rest period of at least 13 consecutive hours between two working periods. Certain exceptions are provided for.

The extensive recourse to shift work in industry has rendered necessary the redefinition of the term "night". Convention No. 90 defines the term as signifying a period of at least 12 consecutive hours, which in the case of young persons under 16 years of age must include the interval between 10 p.m. and 6 a.m., and in the case of young persons between 16 and 18 years of age, an interval prescribed by the competent authority of at least 7 consecutive hours between 10 p.m. and 7 a.m.; different intervals may be prescribed in specific cases.

Non-industrial employment. ILO Convention No. 79 of 1946 concerning the Restriction of Night Work of Children and Young Persons in Non-Industrial Occupations provides that children under 14 years of age shall not be employed nor work at night during a period of at least 14 consecutive hours, including the interval between 8 p.m. and 8 a.m. Room is made for certain exceptions. Young persons between 14 and 18 years of age shall not be employed nor work at night during a period of at least 12 consecutive hours, including the interval between 10 p.m. and 6 a.m. This period may be shortened in countries where the climate renders work by day particularly trying.

ILO Recommendation No. 80 of 1946 on the same subject emphasises the desirability of also adopting appropriate measures for restricting the night work of young persons under 18 years of age who are engaged in domestic service or on work in family undertakings which is not deemed to be harmful, prejudicial or dangerous to adolescents.

Night work by girls. The instruments concerning night work of women apply to all women without distinction as to age, including girls and in particular young women over 18 years of age who are no longer covered by the provisions on night work of children and young persons.

ILO Recommendation No. 13 of 1921 concerning Night Work of Women in Agriculture calls on member States to take steps to regulate the employment of women wage earners in agricultural undertakings during the night, in such a way as to ensure to them a period of rest compatible with their physical necessities and consisting of not less than 9 hours, which shall when possible be consecutive.

ILO Convention No. 89 (Revised) of 1948 concerning Night Work of Women Employed in Industry prohibits the employment of women during the night in industry, except in an undertaking in which only members of the same family are employed. The term "night" as used here signifies a period of at least 11 consecutive hours, including an interval prescribed by the competent authority of at least 7 consecutive hours falling between 10 p.m. and 7 a.m.; the competent authority may prescribe different intervals if necessary, after consulting the employers' and workers' organisations concerned. Certain exceptions are provided for, and the Convention, furthermore, does not apply to women employed in health and welfare services who are not ordinarily engaged in manual work.

Limitations and Prohibitions on the Performance of Certain Kinds of Work by Young Workers

Employment at sea. ILO Convention No. 15 of 1921 fixing the Minimum Age for the Admission of Young Persons to Employment as Trimmers or Stokers provides that young persons under the age of 18 years shall not be employed or work on vessels as trimmers or stokers. An exception is made in the case of young persons of not less than 16 years of age, who, if found physically fit after medical examination, may be so employed on vessels exclusively engaged in the coastal trade of India and Japan. The prohibition does not apply to work done by young persons on school-ships or training-ships, provided that such work is approved and supervised by public authority. A similar waiver is contained in Convention No. 58 (Revised), 1936, fixing the Minimum Age for the Admission of Children to Employment at Sea.

Industrial employment. ILO Convention No. 59 (Revised) of 1937 fixing the Minimum Age for Admission of Children to Industrial Employment provides that in respect of employments which, by their nature or the circumstances in which they are carried on, are dangerous to the life, health or morals of the persons employed therein, national laws shall either prescribe a higher age than 15 years for admission thereto of young persons or adolescents, or empower an appropriate authority to prescribe a higher age than 15 years for such admission.

ILO Recommendation No. 96 of 1953 concerning the Minimum Age of Admission to Work Underground in Coal Mines indicates that young persons under 16 years of age should not be employed underground in coal mines. In addition, young persons who have attained the age of 16 years and are under 18 years of age should not be employed underground in coal mines, except for purposes of apprenticeship, or under conditions governed by the competent authority, after consultation with the employers' and workers' organisations concerned, and relating to the places of work and occupations permitted and the measures of systematic medical and safety supervision to be applied.

ILO Convention No. 123 of 1965 concerning the Minimum Age for Admission to Employment Underground in Mines, which also covers work underground in quarries, provides that persons under a specified minimum age shall not be employed or work underground in mines. The minimum age shall be fixed by national legislation, but shall in no case be less than 16 years.

ILO Recommendation No. 124 of 1965 on the same subject recommends that this minimum age should be progressively raised, with a view to attaining a minimum age of 18 years.

Non-industrial employment. ILO Convention No. 60 (Revised) of 1937 concerning the Age for Admission of Children to Non-Industrial Employment provides that a higher age shall be fixed by national laws or regulations for admission of young persons and adolescents to any employment which, by its nature and the circumstances in which it is to be carried on, is dangerous to the life, health or morals of the persons employed in it.

Special provisions. The prohibitions based on the performance of certain kinds of work by women also extend automatically to girls, and in particular to girls whose age exceeds the minimum ages for admission to employment laid down by the international instruments in respect of children and adolescents.

ILO Convention No. 45 of 1935 concerning the Employment of Women on Underground Work in Mines of All Kinds provides that no female, whatever her age, shall be employed on underground work in any mine. National laws or regulations may exempt from this prohibition females holding positions of management who do not perform manual work, or who are employed in health and welfare services, etc.

Regulation 237 of the ILO Model Code of Safety Regulations for Industrial Establishments provides that the employment of young persons of both sexes under 18 years of age should be prohibited in a series of industrial occupations which are detailed in the Regulations.

ILO Recommendation No. 4 of 1919 concerning the Protection of Women and Children against Lead Poisoning provides that women and young persons over the age of 18 years should be excluded from employment in various processes, a list of which is given. The Recommendation further provides that employment of women and young persons under the age of 18 years in processes involving the use of lead compounds should be permitted only if specific health protection measures are taken.

ILO Convention No. 13 of 1931 concerning the Use of White Lead in Painting prohibits the employment of males under 18 years of age and of all females in any painting work of an industrial character involving the use of white lead or sulphate of lead or other products containing these pigments. An exception is made in the case of painters' apprentices, the protective measures to be taken in such cases being specified.

ILO Convention No. 115 of 1960 concerning the Protection of Workers Against Ionising Radiations applies to all activities involving the exposure of workers to ionising radiations, except when the competent authority, after consulting representatives of employers and workers, considers that its provisions need not be applied owing to the limited radiation doses involved. It provides that no worker under the age of 16 shall be engaged in work involving ionising radiations. Appropriate levels must be fixed for workers who are directly engaged in radiation work and are respectively aged 18 and over, and under the age of 18. Appropriate levels must also be fixed for workers who are not directly engaged in radiation work but who remain or pass where they may be exposed to ionising radiations or radioactive substances. The Convention also specifies various precautions which must be taken.

ILO Convention No. 136 of 1971 concerning Protection Against Hazards of Poisoning Arising From Benzene provides that young persons under 18 years of age shall not be employed in work processes involving exposure to benzene or products containing benzene.

The Committee of Ministers of the Council of Europe has approved the text of a recommendation concerning the minimum age for admission to employment in occupations regarded as dangerous for young persons. A list of such occupations is appended to the recommendation, which provides that the minimum age should generally be fixed at 18 years; however, the competent national authorities may permit young persons under 18 but at all events over 16 years of age to engage in such occupations if this is deemed necessary for their vocational training, and they do so under the supervision and responsibility of a competent person.

Many countries have published lists of work processes or occupations in which young persons may not be employed, and of specific hazards to which they may not be exposed. While some of these prohibitions are common to a number of countries, others exist only in certain countries, and others still are rarely encountered. The age below which these prohibitions apply is generally 18 or 16 years, although it is liable to vary, depending on the type of work concerned.

Sight should also not be lost of the historical reasons which led initially to the adoption of regulations restricting the employment of children and adolescents in dangerous work. Originally, such restrictions or prohibitions were indispensable, because in their absence the children or adolescents concerned, who were still growing, would have been exposed to substantial hazards, at a time when technical safety measures were non-existent or rudimentary, and medical supervision, if any, was highly inadequate. Since then, however, the situation has changed, and the prohibitions on the employment of adolescents in work involving the handling of dangerous substances, for example, are open to re-examination in the light of technological progress, the safety precautions taken, the training received and the possibility of detecting the action of poisons on the organism at an early stage, thanks to adequate medical supervision. This problem should be approached not only by considering what kinds of work young persons should not engaged in, but above all by considering what measures should be taken to make conditions of work sufficiently safe, so that workers in general and young workers in particular may be employed in such work without any risk for life and limb or for their health.

Manual transport of loads. ILO Convention No. 127 concerning the Maximum Permissible Weight to Be Carried by One Worker provides that the assignment of young workers under 18 years of age to manual transport of loads other than light loads shall be limited, and that the maximum weight of such loads shall be substantially less than that permitted for adult male workers. Prior to their assignment to manual transport of loads, workers must receive adequate training or instruction in working techniques, with a view to safeguarding health and preventing accidents.

ILO Recommendation No. 128 on the same subject sets the maximum permissible weight which may be transported manually by an adult male worker at 55 kg. In the case of young workers, the maximum weight should be substantially less than that permitted for adult workers of the same sex. Moreover, young workers under 18 years of age should as far as possible not be assigned to regular manual transport of loads.

Medical Examination of Young Workers

Agriculture. ILO Convention No. 110 of 1958 concerning Conditions of Employment of Plantation Workers provides that every recruited worker shall be medically examined; young workers, therefore, are also covered by this provision. When the journey from the place of recruitment to the place of employment is a long one, the competent authority may require recruited workers to be examined before departure and after arrival at the place of employment. The application of these provisions is obviously of great importance for young workers.

A resolution adopted by the Permanent Agricultural Committee of the ILO emphasises the desirability of young workers under 18 years of age being given a medical examination of fitness for employment before being admitted to employment in agriculture.

Employment at sea. ILO Convention No. 16 of 1921 concerning the Compulsory Medical Examination of Children and Young Persons Employed at Sea provides that the employment of any child or young person under 18 years of age on any vessel shall be conditional on the production of a medical certificate attesting to fitness for such work. The continuation of employment shall be subject to the repetition of such medical examination at intervals of one year, until the young person is 18 years old.

ILO Convention No. 113 of 1959 concerning the Medical Examination of Fishermen provides that no person shall be engaged for employment on a fishing vessel unless he produces a medical certificate attesting to his fitness for the work for which he is to be employed at sea. The nature of the examination is to be determined having due regard to the age of the person concerned and the nature of the duties to be performed. In the case of young persons less than 21 years old, the medical examination must be repeated at annual intervals.

Employment in industry. ILO Convention No. 77 of 1946 concerning Medical Examination for Fitness for Employment in Industry of Children and Young Persons provides that young workers under 18 years of age shall not be admitted to employment by an industrial undertaking unless they have been found fit for the work on which they are to be employed by a thorough medical examination. Fitness for work shall be subject to medical supervision until the young worker is 18 years old, medical examinations being repeated at intervals of not more than one year, or more frequently, when the risks involved in the occupation or the state of health of the child make this desirable. In occupations that involve high health risks medical examinations shall be required until at least the age of 21 years. These examinations must be certified by medical certificates, in the manner laid down by the Convention.

ILO Recommendation No. 79 of 1946 concerning the same subject, but covering non-industrial occupations also, supplements the provisions of Convention No. 77, with special reference to the desirability of extending compulsory medical examination until the age of 21 at least; the scope of such examinations is defined.

ILO Convention No. 124 of 1965 concerning Medical Examination of Young Persons for Fitness for Employment Underground in Mines supplements the provisions of Convention No. 77, the scope of which also includes mines, by requiring yearly medical examinations up to the age of 21 years. An X-ray of the lungs is required at the medical examination and, as necessary, at subsequent examinations.

Non-industrial employment. ILO Convention No. 78 of 1946 concerning Medical Examination of Children and Young Persons for Fitness for Employment in Non-Industrial Occupations provides that children and young persons under 18 years of age shall not be admitted to employment nor work in non-industrial occupations unless they have been found fit for the work in question by a thorough medical examination. Their fitness for such employment must be subject to medical supervision until the age of 18 years, with medical examinations repeated at intervals of not more than one year. As in the case of industrial employment, national laws or regulations must determine in what circumstances re-examination is required at more frequent intervals, and the occupations in which it must be continued until the age of 21 years.

ILO Recommendation No. 79, already referred to, also supplements the provisions of this Convention.
